WHERE DID THAT NUMBER COME FROM?

Chronological Histories and Derivations
of Numbers Important in Science

BRIAN STEDJEE

Copyright © 2021 Brian Stedjee
All rights reserved
First Edition

NEWMAN SPRINGS PUBLISHING
320 Broad Street
Red Bank, NJ 07701

First originally published by Newman Springs Publishing 2021

ISBN 978-1-63692-834-0 (Paperback)
ISBN 978-1-63692-835-7 (Digital)

Printed in the United States of America

CONTENTS

Preface 5

The Length of a Year 9
The Size of the Earth 10
The Diameter of the Moon 11
Orbital Periods of the Planets 13
Relative Distances to the Planets 15
Pi 17
The Speed of Sound 30
The Mass of the Atmosphere 31
e: The base of the Natural Logarithm 32
The Speed of Light 38
Distances to the Sun and Planets 41
0 Degrees, 32 Degrees, and 212 Degrees Fahrenheit 44
The Mass of the Earth 45
The Mass of the Sun 47
Masses of Planets 48
Absolute Zero 49
Distancing to Nearby Stars 50
Atomic and Molecular Weights 51
The Gas Constant R 55
The Mass of the Moon 58
The Rydberg Constant 60
Planck's Constant, the Stefan-Boltzmann Constant, Wien's
Displacement Constant, and Boltzmann's Constant 63
The Surface Temperature of the Sun 67
Avogadro's Number 68
Charge and Mass of an Electron 71

The Size of Atoms 75
Sizes of Nuclei and the Proton 79
Atomic Numbers 82
The Fine Structure Constant 84
The Age of the Earth 87
The Size and Age of the Universe 90
The Temperature of the Earth's Core 93

Bibliography 95

PREFACE

Where did that number come from? Almost everyone has, at one time or another, seen a reported figure for the speed of light or the mass of the sun in an almanac, encyclopedia, or handbook. Those numbers are critical components of the foundations on which the pure and applied sciences, such as chemistry, physics, astronomy, earth sciences, biology, and engineering, are built. Yet, most people probably take them as matters of fact without ever considering their sources.

Over the centuries, some known and countless unknown researchers, driven by curiosity about and fascination with the Earth and what lies beyond it, used their creativity and imaginations to find the special numbers that are such critical components of today's complex technologies. The results of their work are seen everywhere in our lives today; yet most people have never heard their names.

This book presents, in chronological order, a series of brief historical sketches that, hopefully, will help explain to the educated layperson where those amazing numbers come from. It is possible that some people simply don't believe in the reasoning behind the discoveries and may ask, "How could anybody figure something like that out anyway?"

If a person wanted to answer that question—really wanted to find out where a particular number came from—he could simply call the local college and ask a teacher, couldn't he? He could call, but he might not get an answer. Many teachers simply do not know the origins of the numbers they memorize and solve problems with, nor do all teachers realize that the numbers change over time as more and more sensitive experiments are performed.

It was the realization of my own ignorance of the histories of many (if not most) of the numbers we accept, without questioning that prompted the research into the discovery and, in many cases, evolution of the numbers over time with generations of researchers basing their work on the results of experiments and analyses done by their predecessors.

For example, most of the values for masses and distances of objects are based on experimentally determined constants. The mass of the Earth calculation depends on the universal gravitational constant G. The mass of an electron calculation requires the elementary charge constant e. And to find the distance to Mars, you need the speed of light C.

The values for these constants are not the same in the newer handbooks as they are in the older ones because the experiments have improved. This is important because mass and distance values cannot be known to any greater accuracy than that of the constants they are based on.

Here are some examples of numbers whose accuracy has improved over time as researchers built on and learned from the work done by others, often centuries earlier:

Constant	Symbol	Discoverer	Year of Discovery
Speed of light	C	Roemer	1666
Gravitational	G	Cavendish	1798
Avogadro's number	N	Planck	1900
		Millikan	1910
Charge of an electron	e	Millikan	1910

The degree of accuracy of some of those constants has improved considerably over the years.

Constant	1926 value	1968 value	1983 value	2014 value
c	2.9986×10^8 meters/second	2.997925×10^8 meters/second	2.99792458×10^8 meters/second	Research results announced in 2013 suggest this may not be a constant.
G	6.66×10^{-11} nt. meter^2/kg^2	6.670×10^{-11} nt. meter^2/kg^2	6.6720×10^{-11} nt. meter^2/kg^2	6.6738×10^{-11} nt. meter^2/kg^2

N	6.061 × 10²³ particles/mole	6.0225 × 10²³ particles/mole	6.022043 × 10²³ particles/mole	6.02214129 × 10²³ particles/mole
e*	1.59 × 10⁻¹⁹ coulombs	1.6021 × 10⁻¹⁹ coulombs	1.6021892 × 10⁻¹⁹ coulombs	1.602176565(35) × 10⁻¹⁹ coulombs

*This is the elementary charge constant, which should not be confused with *e*—the natural logarithm.

THE LENGTH OF A YEAR

No one knows who first determined the length of a year. We do know that the people who built Stonehenge (ca. 2000 BC) obviously knew it, as well as a number of other astronomical facts. Because the Earth is tilted on its axis, the Sun appears to rise and set farther and farther to the north (in the northern hemisphere) until June 21 each year.

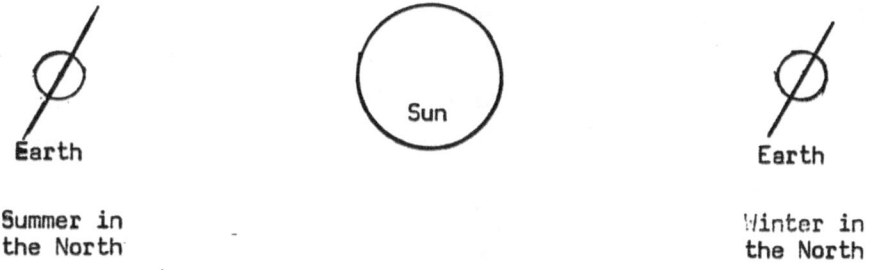

Figure 1-1

Then it begins rising and setting farther south. Given two set points to sight the sunrise, it's possible to determine June 21 each year.

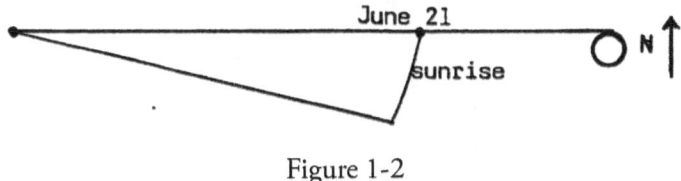

Figure 1-2

The length of a year is 365.256 days.

THE SIZE OF THE EARTH

The Greeks knew that the Earth is spherical and one of them, Eratosthenes (276–195 BC), measured it. Eratosthenes noticed that on June 22 in Syene, Egypt (near modern day Aswan), a sundial cast no shadow. Another version of this story tells us that Eratosthenes noticed that sunlight struck the bottom of a vertical well at that time of day. At Alexandrea, Egypt (500 miles north of Syene), sunlight arrives 7 1/5 degrees from the vertical at noon on June 22. Eratosthenes reasoned that if the sun's rays were parallel and the Earth was spherical, then the ratio between 7 1/5 degrees and the number of degrees in a full circle (360) multiplied by the distance between Alexandria and Syene should give the circumference of Earth.

$$\frac{360 \text{ degrees} \times 500 \text{ miles}}{7\frac{1}{5} \text{ degrees}} = 25{,}000 \text{ miles}$$

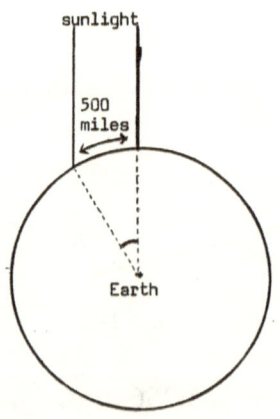

Figure 2-1

The circumference of Earth at the equator is, in fact, 24,902 miles or 40,077 kilometers.

THE DIAMETER OF THE MOON

Sometime between 160 and 130 BC, the Greek astronomer, Hipparchus, successfully determined the size of the Moon and its distance from Earth. He based the calculation on the size of Earth's shadow on the Moon during a lunar eclipse. Because the sun is the same angular width as the Moon, Earth's shadow on the Moon is one lunar diameter smaller than the diameter of Earth itself. Earth's shadow is 8/3 the diameter of the Moon. This may be found by taking, during the longest lunar eclipses, the ratio of the time the leading edge of Earth's shadow crosses the moon to the time the following edge begins to appear.

Since Earth's shadow is 8/3 the diameter of the Moon, Earth itself must be 8/3 + 1 = 11/3 the diameter of the Moon. The diameter of Earth is 12,756 kilometers. So using this method, the diameter of the Moon is calculated to be 3/11 multiplied by the diameter of Earth (3/11 × 12,576 km = 3479 km). The Moon's actual diameter is 3,476 kilometers or 2,160 miles.

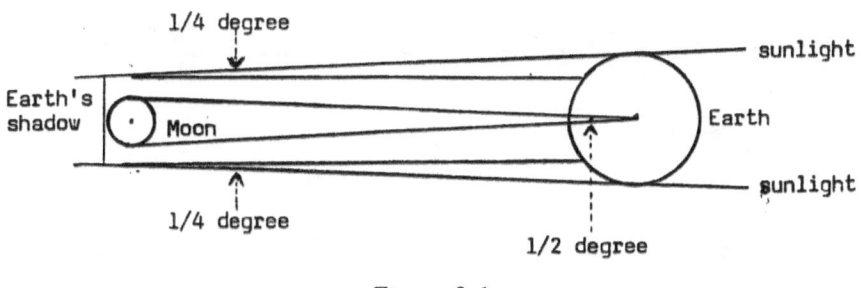

Figure 3-1

Knowing the diameter of the Moon and its angular width in the sky, Hipparchus was able to calculate its distance from Earth using triangulation.

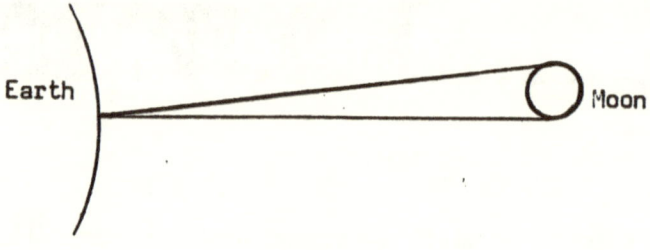

Fig 3-2

Around the year AD 140, the Greek astronomer Ptolemy found an alternative method of measuring the distance to the Moon. The distance can be found by parallax if two widely separated observers simultaneously sight a spot on the Moon. However, in Ptolemy's day, when radio communication and accurate clocks were scarce, the idea seemed somewhat impractical. Triangulation can be done, though, by finding the difference a point on the Moon varies from shifting 90 degrees as the earth rotates 90 degrees (1/4 day), although the angle the Moon makes in its orbit around Earth during 1/4 day must be subtracted from the observed angular difference. Using this method, Ptolemy determined the moon's distance from Earth to be 59 Earth diameters.

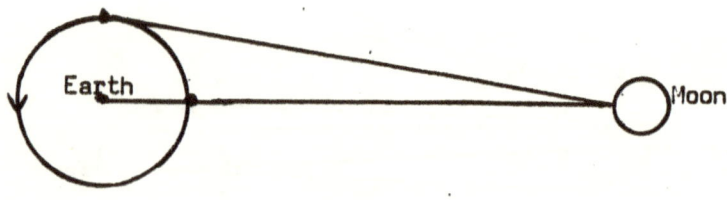

Figure 3-3

Nowadays the distance to the Moon can be found very accurately by bouncing a radar signal off of it. Knowing the speed of light and the time it takes for the signal to return, the distance to the Moon can be determined to within meters. The average distance to the Moon is 384,400 kilometers or 238,857 miles. The distance actually varies between 363,300 kilometers and 405,500 kilometers.

ORBITAL PERIODS OF THE PLANETS

Nicolaus Copernicus was not the first to theorize that the planets orbit around the Sun, but he was the first to calculate their orbital periods. The method is different for the inner planets than it is for the outer planets. Consider the inner planet Venus.

Earth travels at a rate of 360 degrees/year around the sun. Therefore, between one conjunction with Venus and the next, Earth makes an arc following this equation: (360 degrees/year)(the time between conjunctions). Venus travels at a rate of 360 degrees/Venus year. But Venus makes an entire orbit before a second conjunction may be observed. So the number of degrees Venus travels between conjunctions is (360 degrees/Earth year)(the time between conjunctions + 360 degrees).

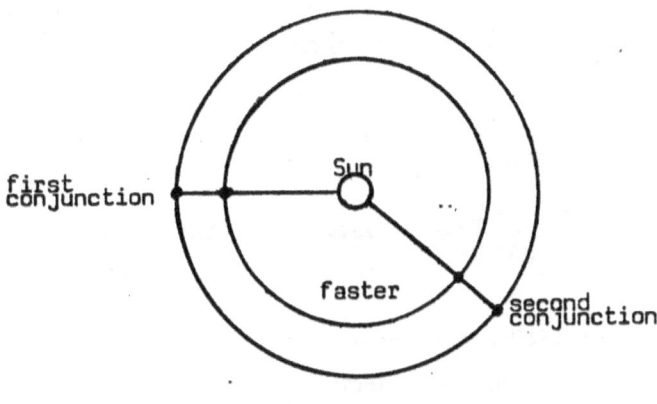

Figure 4-1

Therefore,

$$\left(\frac{360 \text{ degrees}}{\text{Earth year}}\right)(\text{time between conjunctions}) + 360 \text{ degrees} =$$
$$\left(\frac{360 \text{ degrees}}{\text{Venus year}}\right)(\text{time between conjunctions})$$

Dividing both sides of the equation by 360 degrees, you get

$$\left(\frac{1}{\text{Earth year}}\right)(\text{time between conjunctions}) + 1 =$$
$$\left(\frac{1}{\text{Venus year}}\right)(\text{time between conjunctions})$$

The only unknown is the Venus year, which now may be calculated in terms of Earth years.

In the case of the outer planets, it is Earth that goes more than a full orbit between conjunctions. So the equation for finding an outer planet's period is correspondingly altered.

$$\left(\frac{1}{\text{Mars year}}\right)(\text{time between conjunctions}) =$$
$$\left(\frac{1}{\text{Earth year}}\right)(\text{time between conjunctions}) - 1$$

The equations for finding orbital periods for planets are usually simplified even further. The general forms are

$$\frac{1}{\text{Venus year}} = 1 + \left(\frac{1}{\text{time between conjunctions}}\right)$$
$$\frac{1}{\text{Mars year}} = 1 - \left(\frac{1}{\text{time between conjunctions}}\right)$$

RELATIVE DISTANCES TO THE PLANETS

Besides calculating the orbital periods of the planets, Copernicus found their relative distances from the Sun. The operation is easier with the inner planets than it is with the outer ones. When an inner planet, such as Venus, appears as far away from the Sun as possible, it forms a right triangle with the Sun and Earth.

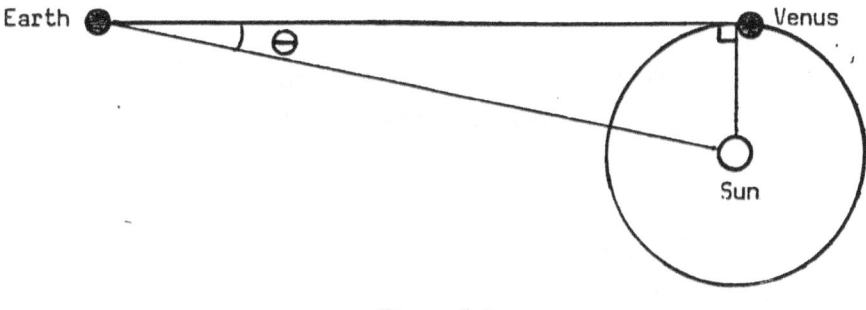

Figure 5-1

So the distance Venus orbits from the Sun is sine θ times Earth's distance from the Sun.

In the case of outer planets, positions in orbit can be calculated using their orbital velocities. If the position is calculated when the planet and the Sun are observed at 90 degrees from one another, the angle the planet forms with respect to Earth and the Sun may be determined, and hence, so may its relative distance from the Sun.

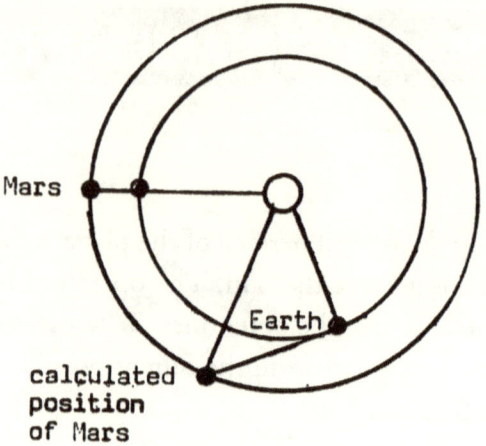

Figure 5-2

However, Copernicus did not know the absolute distance to any of the planets because the absolute distance to the Sun was not known.

PI

Pi is a remarkable number in that it is involved in calculating both the area (πr^2) and the perimeter ($2\pi r$) of a circle. In ancient times, pi was believed to be either 3 or 3 1/8. These figures were apparently found by rather crude measurements. The 3 1/8 figure is from the Rhind Papyrus (ca. 1500 BC) and the 3 figure is implied in the Bible (1 Kings 7:23).

There is evidence that some ancient peoples either wanted pi to be equal to 3 or placed some religious significance on the number 3. Several imperfect stone circles built by pre-Celtic people in Northern Europe have been investigated, and it is interesting that they are all imperfect in the same way. Measurements reveal that the ratios of the circumferences of the "circles" to their narrowest diameters is almost always exactly 3.

The first mathematical attempts to calculate pi were apparently made by the Greeks. Although Archimedes is usually given credit for the first calculation, he apparently had two predecessors, Antiphon and Bryson of Heraclea. Antiphon suggested finding the value of pi by drawing polygons of increasing numbers of sides inside a circle, calculating the length of each side, then multiplying by the total number of sides. Bryson of Heraclea suggested drawing polygons outside the circle. Polygons drawn inside a circle always give values of pi slightly less than the real value, and polygons drawn outside the circle always give values slightly greater than the real pi. Archimedes was able to narrow in on the true value of pi by combining both methods.

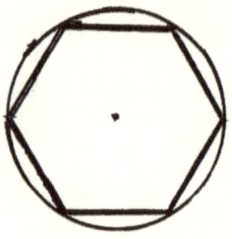 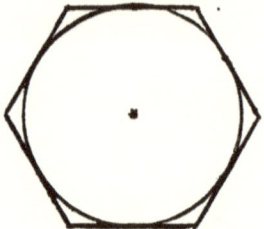

Figure 6-1

Using the trigonometric functions we have today, it is possible, without great difficulty, to obtain a fairly good value of pi using the polygon method. However, when ancient Greeks were doing their calculating, things like sines and tangents were as yet undefined (at least in the modern sense), the calculations were extremely tedious.

Given a square inscribed in a circle with a radius of 1, the perimeter of the square may be found using the Pythagorean theorem.

$$A^2 + B^2 = C^2 \text{ or } \left(\sqrt{\tfrac{1}{2}}\right)^2 + \left(\sqrt{\tfrac{1}{2}}\right)^2 = 1^2$$

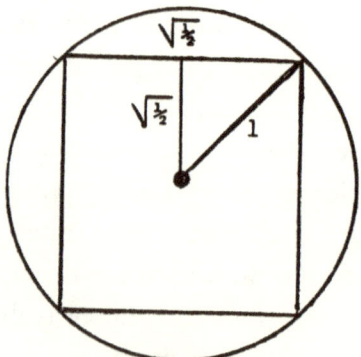

Figure 6-2

So half of each side of the square has a length of the square root of A, and each side has a length of two times the square root of A. The perimeter of the square is four times twice the square root of 1/2 =

5.657. This compares with a circle circumference of 6.283. Doubling the number of sides gives a more accurate value.

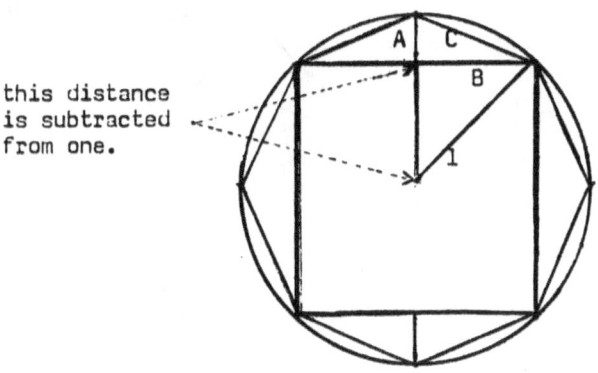

Figure 6-3

The B sides of the small triangles around the perimeter have already been determined to be the square root of A. The A sides of the small triangles are the radius (1) minus the distance from the bases of the small triangles to the center of the circle (square root of 1/2) or 1 minus the square root of 1/2. Each side C of the resulting octagon then is

$$\sqrt{\left(\sqrt{\tfrac{1}{2}}\right)^2 + \left(1 - \sqrt{\tfrac{1}{2}}\right)^2} = \sqrt{0.5 + 0.08579} = 0.76537$$

The perimeter is 8 (.76537) = 6.1229. The number of sides of the octagon can be doubled again to give a polygon of sixteen sides and a more accurate value for pi.

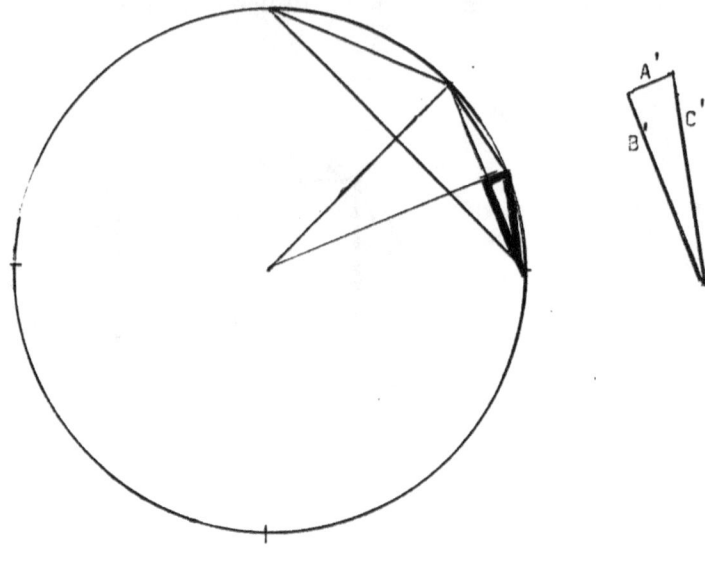

Figure 6-4

$$\left[\frac{\sqrt{\left(\sqrt{\frac{1}{2}}\right)+\left(1-\sqrt{\frac{1}{2}}\right)^2}}{2}\right]^2 + \left[1 - \sqrt{\left(\sqrt{\frac{1}{2}}\right)^2 + \left(1-\sqrt{\frac{1}{2}}\right)^2}\right]^2 = c^2$$

$$(B^1)^2 + (A^1) = c^2$$

Here, *C* is the length of one side of the new polygon. This process of doubling the sides of a polygon can be repeated again and again, but as can be seen, the equations soon become extremely cumbersome.

When Archimedes worked on pi, he started with hexagons. Using our trigonometry today, it's pretty easy to find the perimeters of polygons. When the hexagon is drawn around the outside of the circle, the angle θ is 360 degrees/12 = 180 degrees/6 = 30 degrees, and the length of side *B* is $\frac{\text{side B}}{\text{side A}} = \frac{\text{side B}}{1}$ = the tangent of 30 degrees.

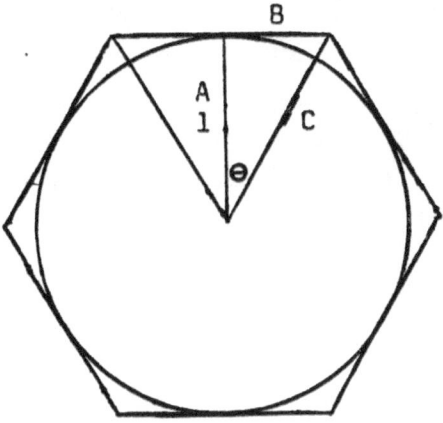

Figure 6-5

The perimeter of this hexagon then is 12B = 12 (tangent of 30 degrees) = 3.464. As the number of sides of the polygon is increased, the perimeter is always the number of sides times tangent (180 degrees/ the number of sides).

When a hexagon is drawn on the inside of the circle, the length of side B is $\frac{\text{side B}}{\text{side C}} = \frac{\text{side B}}{1}$ = the sine of 30 degrees.

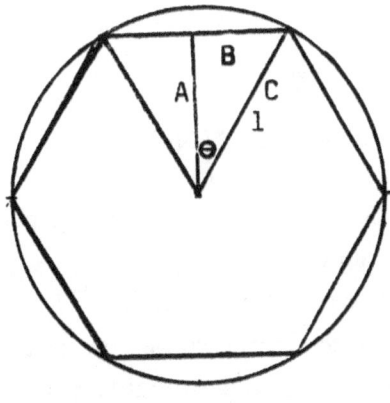

Figure 6-6

So the perimeter of this hexagon is 12 times the sine of 30 degrees = 3.0.

Archimedes increased the number of sides of polygons drawn both inside and outside a circle until he narrowed the value of pi down to between 3.141 and 3.142.

The first modern person to determine pi using something similar to the modern approach was French mathematician, François Viète. In 1592, he found that

$$\frac{1}{\sqrt{\frac{1}{2}} \sqrt{\frac{1}{2}+\frac{1}{2}\sqrt{\frac{1}{2}}} \sqrt{\frac{1}{2}+\frac{1}{2}\sqrt{\frac{1}{2}+\frac{1}{2}\sqrt{\frac{1}{2}}}}} = \frac{\pi}{2}$$

The formula is known as Viète's infinite product, and as the name implies, pattern of the formula continues infinitely. Written as follows, its pattern can also be written as

$$\frac{1}{\sqrt{\frac{1}{2}}} \times \frac{1}{\sqrt{\frac{1}{2}+\frac{1}{2}\sqrt{\frac{1}{2}}}} \times \frac{1}{\sqrt{\frac{1}{2}+\frac{1}{2}\sqrt{\frac{1}{2}+\frac{1}{2}\sqrt{\frac{1}{2}}}}} \times \ldots = \frac{\pi}{2}$$

Somewhat similar to the Greek method, this infinite product is derived by observing changes that happen when one doubles the number of sides of polygons inscribed in a circle. In the case of the infinite product, however, it is the area rather than the perimeter of the circle that is used to calculate pi.

The circumference of a circle (Figure 6-7), two π times its radius, is also 360 degrees. Consequently, 90 degrees and $\frac{\pi}{2}$ may be considered equal angles. The central angle of a triangular portion of a polygon then may be written in terms of pi as well as degrees. The central angle of the triangular portion of a square inscribed in a circle is $\frac{360 \text{ degress}}{4} = 90$ degrees or $\frac{2\pi}{4} = \frac{\pi}{2}$.

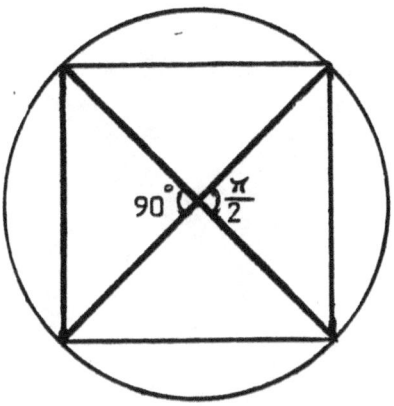

Figure 6-7

In general, the central angle of any triangular portion of a polygon composed of identical triangles is $\frac{2\pi}{N}$ where N is the number of sides the polygon has.

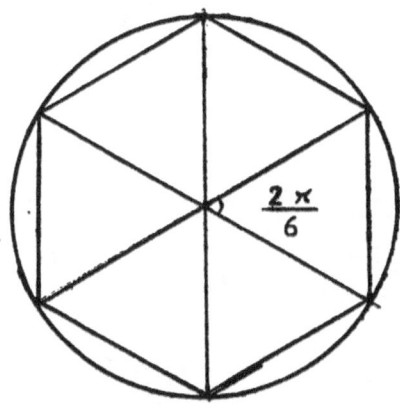

Figure 6-8

When the number of sides of an inscribed in a square is doubled, the areas of the triangles of the resulting octagon can be calculated using the measurements of the original triangles.

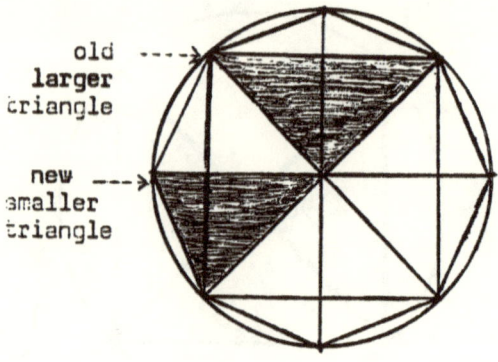

Figure 6-9

In the following figure, the sides of one of the larger triangles are *AB*, *BD*, and *DA*. The two smaller triangles are *AB*, *BC*, *CA* and *AC*, *CD*, and *DA*.

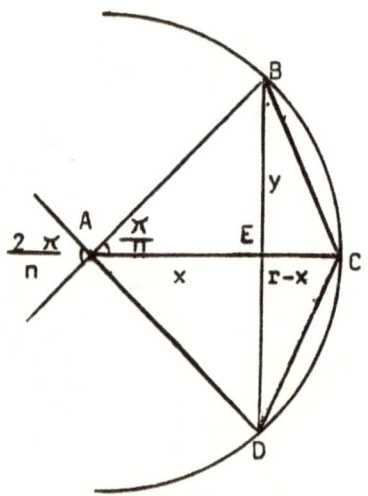

Figure 6-10

The area of the larger triangle *ABD* is *A* its base from *B* to *D* times its height: $\frac{1}{2}(2y)(x) = yx$. The areas of the new smaller triangles *ABC* and *ADC* are $\frac{1}{2}$ their bases *AC* times their heights, *EB* for one and *ED* for the other.

24 | BRIAN STEDJEE

The base AC is $x + (r-x)$ where r is the radius of our circle, and the height EB is the distance y. The area of the smaller triangle then is $1/2y(x + r - x)$, which simplifies to yr.

$$\frac{\text{the area of the smaller triangle}}{\text{the area of the larger triangle}} = \frac{1yr}{2yx} = \frac{r}{2x}$$

In figure 6-10, points A, B, and E form a right triangle. Side x is defined by points A and E, side y by points B and E, and the hypotenuse of the triangle by points A and B. The angle formed by side x and the hypotenuse of the triangle, which in the figure is also the radius of the circle r is $\frac{\pi}{N}$.

We know from trigonometry that the cosine of that angle is $\frac{x}{r}$ and that the reciprocal of the cosine, the secant, is $\frac{r}{x} = \frac{\text{hyptenuse}}{\text{adjacent}} =$ secant of $\frac{\pi}{N}$

Thus, the area of the smaller triangle is $\frac{1}{2}$ the secant of $\frac{\pi}{N}$ multiplied by the area of the larger triangle. The number of triangles in the octagon is double the number of triangles that make up the square, so the area of the octagon is 2 ($\frac{1}{2}$ the secant of $\frac{\pi}{N}$) (the area of the square), which is the secant of $\frac{\pi}{N}$ multiplied by the area of the square.

The area of a square drawn inside a unit circle (a circle with a radius of 1) is 2. Each side of the square is $2\sqrt{\frac{1}{2}}$ so, $2\sqrt{\frac{1}{2}} \left(2\sqrt{\frac{1}{2}} \right) = 4 \times \frac{1}{2} = 2$

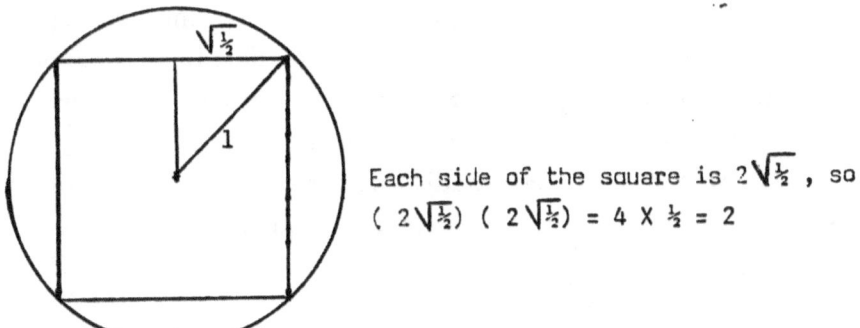

Each side of the square is $2\sqrt{\frac{1}{2}}$, so $(2\sqrt{\frac{1}{2}})(2\sqrt{\frac{1}{2}}) = 4 \times \frac{1}{2} = 2$

Figure 6-11

Changing from a square to an octagon, by doubling the number of sides of the figure, results in an octagon with an area that equals the area of the square times the secant of $\frac{\pi}{4}$. This equals 2 (secant of $\frac{\pi}{4}$). The secant of $\frac{\pi}{4}$ (or 45 degrees) is the square root of 2 or $\frac{1}{\sqrt{\frac{1}{2}}}$ so the area of the octagon is $2 \frac{1}{\sqrt{\frac{1}{2}}}$.

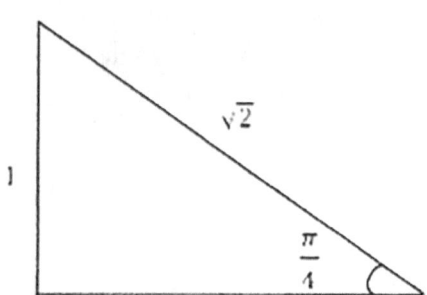

$$\text{secant} = \frac{\text{hyptenuse}}{\text{adjacent}} = \frac{\sqrt{2}}{1} = \sqrt{2} = \frac{1}{\sqrt{\frac{1}{2}}}$$

The cosine or adjacent hypotenuse is $\frac{1}{2}$.

When the sides of the octagon are doubled to 16, the area of the 16-sided polygon is the area of the octagon multiplied by $2\frac{1}{\sqrt{\frac{1}{2}}}$ multiplied by the secant of $\frac{\pi}{8}$. This results from the fact that whenever the number of sides of an inscribed polygon is doubled, the resulting area is (secant $\frac{\pi}{N}$)(the area of the original polygon). This is what happened when the square was expanded to an octagon. It is what happens when the octagon is expanded to a hexadecagon, etc.

The secant (hypotenuse/adjacent) is the reciprocal of the cosine, and there is a trigonometric equation that says that

$$\frac{\text{the cosine of an angle}}{2} = \sqrt{\frac{1+\text{cosine of the angle}}{2}}$$

Consequently,

$$\text{the secant of } \frac{\pi}{8} = \frac{1}{\text{cosine } \frac{\pi}{8}} = \frac{1}{\sqrt{\frac{1}{2} + \frac{1}{2}\text{cosine } \frac{\pi}{8}}} = \frac{1}{\sqrt{\frac{1}{2} + \frac{1}{2}\sqrt{\frac{1}{2}}}}$$

So the area of the hexagon inscribed in our circle with a radius of 1 is

$$2 \times \cfrac{1}{\sqrt{\frac{1}{2}}\sqrt{\frac{1}{2}+\frac{1}{2}\sqrt{\frac{1}{2}}}}$$

Recalling Viète's infinite product and its first three terms,

$$\cfrac{1}{\sqrt{\frac{1}{2}}\sqrt{\frac{1}{2}+\frac{1}{2}\sqrt{\frac{1}{2}}}\sqrt{\frac{1}{2}+\frac{1}{2}\sqrt{\frac{1}{2}+\frac{1}{2}\sqrt{\frac{1}{2}}}}}$$

or

$$\cfrac{1}{\sqrt{\frac{1}{2}}} \times \cfrac{1}{\sqrt{\frac{1}{2}+\frac{1}{2}\sqrt{\frac{1}{2}}}} \times \cfrac{1}{\sqrt{\frac{1}{2}+\frac{1}{2}\sqrt{\frac{1}{2}+\frac{1}{2}\sqrt{\frac{1}{2}}}}} \cdots = \frac{\pi}{2}$$

You may recognize that the area of the octagon $2\,\cfrac{1}{\sqrt{\frac{1}{2}}}$ is exactly twice the first term in the above formula $\cfrac{1}{\sqrt{2\frac{1}{2}}}$ and that the area of the hexadecagon is exactly twice the product of the first and second terms of Viète's formula: $2\,\cfrac{1}{\sqrt{\frac{1}{2}}\sqrt{\frac{1}{2}+\frac{1}{2}\sqrt{\frac{1}{2}}}}$

Furthermore, if the number of sides of the polygon were doubled once again to 32, the first three terms of Viète's formula would be used to calculate the area of the 32-sided polygon:

$$2 \times \cfrac{1}{\sqrt{\tfrac{1}{2}} \; \sqrt{\tfrac{1}{2}+\tfrac{1}{2}\sqrt{\tfrac{1}{2}}} \; \sqrt{\tfrac{1}{2}+\tfrac{1}{2}\sqrt{\tfrac{1}{2}+\tfrac{1}{2}\sqrt{\tfrac{1}{2}}}}}$$

This is because the more sides the inscribed polygon has in a circle with a radius of 1, the closer the total area approaches pi. That is, pi = (the area of the square) (secant $\tfrac{\pi}{4}$) (the area of the octagon) (secant $\tfrac{\pi}{8}$) (the area of the hexadecagon) (secant $\tfrac{\pi}{16}$), etc.

Pi also equals

$$2 \times \cfrac{1}{\sqrt{\tfrac{1}{2}} \; \sqrt{\tfrac{1}{2}+\tfrac{1}{2}\sqrt{\tfrac{1}{2}}} \; \sqrt{\tfrac{1}{2}+\tfrac{1}{2}\sqrt{\tfrac{1}{2}+\tfrac{1}{2}\sqrt{\tfrac{1}{2}}}}} \ldots$$

Dividing it by 2 gives Viète's formula:

$$\frac{\pi}{2} = \cfrac{1}{\sqrt{\tfrac{1}{2}} \; \sqrt{\tfrac{1}{2}+\tfrac{1}{2}\sqrt{\tfrac{1}{2}}} \; \sqrt{\tfrac{1}{2}+\tfrac{1}{2}\sqrt{\tfrac{1}{2}+\tfrac{1}{2}\sqrt{\tfrac{1}{2}}}}}$$

The advent of calculus in the 1600s resulted in the discovery of several infinite series involving pi. Ratios or functions of pi often result when one applies the methods of calculus to the equation of a circle. Viète's formula was an infinite product. The first infinite sum involving pi was discovered in 1674 by Gottfried Leibniz. The equation Leibniz discovered is $\frac{\pi}{4} = 1 - \frac{1}{3} + \frac{1}{5} - \frac{1}{7} + \frac{1}{9} - \frac{1}{11} \ldots$

This infinite series results from a Maclaurin series expansion that says that the arctangent or angle whose tangent is x equals $x - \frac{x^3}{3} + \frac{x^5}{5} + \frac{x^7}{7} + \frac{x^9}{9} \ldots$

The angle whose tangent is 1 is 45 degrees or $\frac{\pi}{4}$ and so the angle whose tangent is 1 equals $\frac{\pi}{4} = 1 - \frac{1}{3} + \frac{1}{5} - \frac{1}{7} + \frac{1}{9} \ldots$

It is an interesting fact that neither the name pi nor the symbol π we know today was known (except as a letter of the Greek alphabet) before 1706, when English mathematician, William Jones, suggested its use. It was later popularized by one of the outstanding mathematicians of the eighteenth century, Leonard Euler.

Utilizing infinite series more convenient than Viète's or Leibniz's formulas, the value of pi has been worked out to 100 million decimal places. All of them will not be listed here, but the first 15 are 3.14159265358979.

THE SPEED OF SOUND

It's not known with certainty who first measured the speed of sound. The first person known to do so, Marin Mersenne, obtained a value of 1,380 feet per second. He did so by timing the sound from pistol shots at known distances. He was able to detect when the pistols were fired from the flashes they emitted. Other early experimenters were Pierre Gassendi (1,473 ft/sec), Giovanni Cassini, Olaus Roemer, Jean Picard, and Christian Huygens of the Paris Academy (1,172 ft/sec.).

The speed of sound, however, varies with the gas transmitting it and its temperature. In his *Principia*, Isaac Newton published an equation that gave an approximate velocity of sound. A more accurate equation was suggested by Pierre-Simon Laplace in 1816. The expression is

$$\text{velocity} = \sqrt{\frac{\text{a constant} \times \text{pressure}}{\text{density}}}$$

The speed of sound in air is 1,129 ft/sec or 344 m/sec at 20 degrees centigrade.

THE MASS OF THE ATMOSPHERE

In 1648, Blaise Pascal calculated and published the mass of the atmosphere in his *Récit de la grande expérience de l'équilibre des liqueurs*. Pascal wasn't particularly proud of the discovery. He stated that "a child who knew how to add and subtract could do it." The calculation actually is relatively simple. The average air pressure at the surface of Earth is 14.7 pounds per square inch. That means that if all the air in a column is 1 inch square—stretching from the surface of Earth into space—were condensed and weighed, it would weigh 14.7 pounds. So the weight of the atmosphere is simply 14.7 pounds per square inch multiplied by the number of square inches on Earth.

The radius of the Earth is

$$\left(3960 \text{ miles} \times 5280 \frac{\text{ft}}{\text{mile}}\right)\left(12 \frac{\text{inches}}{\text{foot}}\right) = 2.5 \times 10^8 \text{ inches}$$

So its surface area is $4\pi (2.5 \times 10^8 \text{ inches})^2 = 7.8 \times 10^{17}$ square inches.

Therefore, the atmosphere weighs $\left(14.7 \frac{\text{lbs}}{\text{square inch}}\right)$ $(7.8 \times 10^{17} \text{ square inches}) = 1.1 \times 10^{19}$ pounds - or 5.2×10^{18} kilograms.

E: THE BASE OF THE NATURAL LOGARITHM

In 1614, Scottish mathematician John Napier published a paper on trigonometry. In the paper, Napier introduced a new definition he called logarithm a concept that facilitated the calculation of quantities that are decreasing continually when the rate of that decrease depends on the quantity remaining. Henry Briggs at Cambridge found in 1624 that calculations using logarithms in our base 10 mathematical system were simplified when the logarithms also had a base of 10 (i.e., log of 10 = 1).

A logarithm is actually an exponent, and an exponent can have any base. For example, another way of saying $2^3 = 8$ is to say that 3 is the log of 8 in base 2. Actually, there are only two logarithmic bases that are commonly used. They are base-ten and base e (which will be explained later).

When powers of ten are written in exponential notation, they are multiplied by adding the exponents (e.g., $1 \times 10^3 \times 10^5 = 1 \times 10^8$). To divide, the exponent of the divisor is subtracted from the exponent of the dividend (e.g., $\frac{1 \times 10^9}{1 \times 10^2} = 1 \times 10^7$). Since logarithms (often referred to simply as logs) are exponents, numbers can be multiplied by adding logs or dividing 3 + 0 =3.

It is also true of logs, in any base, that the log of the base is equal to 1.

($2^1 = 2$, $10^1 = 10$, and $e^1 = e$)

Through the use of log tables, logs can also facilitate multiplying and dividing numbers that aren't multiples of 10. Logs (exponents) are numbers written as powers of 10. Just as $1{,}000 = 1 \times 10^3$, and $.001 = 1 \times 10^{-3}$, $2 = 1 \times 10^{.301}$ and $50 = 1 \times 10^{1.699}$. The only difference is that the exponents, or logs, from tables are written in decimal form rather than as whole numbers. The log of 2 (.301) plus the log of 50 (1.699) = 2.000, which is the log of 100 or 50 times 2.

Some general properties of logarithms (regardless of base) are

1. log of 1 = 0
2. log of the base = 1
3. log of A times B = log of A + log of B. For example, log of 1,000 times 100 = 3 + 2 = 5
4. log of A/B = log of A − log of B. For example, log of 100/.1 = 2 − (−1) = 3
5. log of $(A)^B$ = B times log of A. For example, log of 100^3 = 3 times 2 = 6

Mathematicians in the 1600s were interested in finding out more about logs than just their general properties. In the 1660s, Isaac Newton pioneered a method of finding the slope of a curve on a graph. The slope of a line is the "rise divided by the run" or the y component divided by the x component. On a graph, the value of y usually depends on the value of x. Y is said to be a function of x, or in symbolic form, y = f(x). Newton found that he could isolate a tiny portion of the graph and find the slope of the curve at that point in the following manner. Slope is (change of y)/(change of x), but y depends on x, so the general slope of a function is $\frac{\text{change of } y}{\text{change of } x} = \frac{\text{change of a function of } x}{\text{change of } x}$

To find the slope of a curve over a small portion of the graph, the procedure is to change the above expression to $\frac{\text{change of } y}{\text{change of } x} = \frac{\text{change of function of }(x + \text{a tiny increment of } x) - \text{change of function of } x}{\text{change of a tiny increment of } x}$

This is called the derivative of the function and may be written symbolically as $\frac{\text{change of } y}{\text{change of } x} = \frac{dy}{dx}$ = the limit of $\frac{f(x + \Delta x) - f(x)}{\Delta x}$ as Δx goes to zero. Both d and Δ mean "a little change."

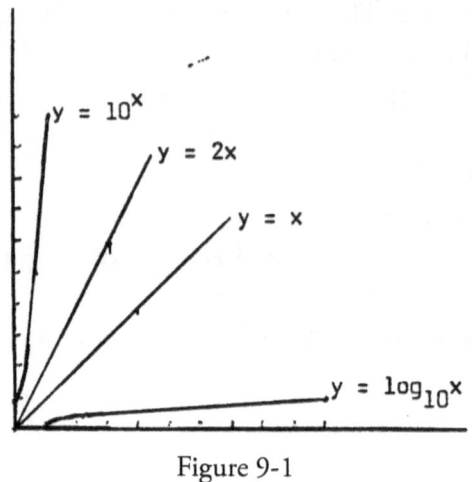

Figure 9-1

To find the slope (or derivative) of the curve of the formula y = log of x, the log is the function of *x*, so

$$\frac{\text{change of y}}{\text{change of x}} = \frac{\log(x + \text{a tiny increment of } x) - \log(x)}{\text{the tiny increment of } x}$$

When the slope of the log function is found using the assumption above, a surprising fact is revealed. There is an ideal or natural logarithm whose base has been known since 1665, when Isaac Newton discovered it. The name of that base is *e*, however, it has only been around since the middle 1700s when Swiss mathematician Leonard Euler named it.

When *e* is raised to any power, the slope of the curve at any point is equal to the equation of the curve itself. This is written symbolically as, $\frac{de^x}{dx} = e^x$. The following process leads naturally to *e*. Remember that $\frac{d \log x}{dx}$ = slope of logx = limit of $\frac{\log(x + \Delta x) - \log(x)}{\Delta x}$. Another way to say this is slope of log x = limit $\frac{1}{\Delta x}$ times the log of (x + Δx) - log(x). When the log of one number is subtracted from the log of another number, the result is the log of the quotient. This is from logarithm property 4, which is LogA - LogB = Log$\frac{A}{B}$ So another way of writing log(x + Δx) -

$\log(x)$ is $\log \frac{x+\Delta x}{x}$. The derivative (slope) of the natural log can therefore be written as slope or $d \log(x) = \text{limit} \frac{1}{\Delta x} \log \left(\frac{x+\Delta x}{x}\right)$ as Δx goes to zero.

Any number multiplied by 1 is itself. When you multiply the above expression by 1 in the form $\frac{x}{x}$ the expression $d \log x = \text{limit} \frac{x}{x \Delta x} \log \left(\frac{x+\Delta x}{x}\right)$ results. The log of a number multiplied by some number, B is the same as the log of that number raised to the B power. This is from logarithm property 5, which is $\log A$ times $B = \log A^B$. For example, 3 times $\log 100 = \log 100^3$.

In the expression that gives the slope or derivative of the natural log of x, the log of $\left(\frac{x+\Delta x}{x}\right)$ is multiplied by $\left(\frac{x}{x\Delta x}\right)$ or $\frac{1}{x} \frac{x}{\Delta x}$

Since multiplying a log by a number is the same as raising it to a power, $\frac{x}{\Delta x} \log \left(\frac{x+\Delta x}{x}\right)$ is the same as $\log \left(\frac{x+\Delta x}{x}\right)^{\frac{x}{\Delta x}}$ Therefore, the derivative, $\text{limit} \frac{1}{x} \frac{x}{\Delta x} \log \left(\frac{x+\Delta x}{x}\right) = \text{limit} \frac{1}{x} \log \left(\frac{x+\Delta x}{x}\right)^{\frac{x}{\Delta x}}$ or, $\frac{1}{x} \log \left(x + \frac{\Delta x}{x}\right)^{\frac{x}{\Delta x}}$

In the expression $\left(x + \frac{\Delta x}{x}\right)^{\frac{x}{\Delta x}}$, the term $\frac{\Delta x}{x}$ inside the parentheses is the reciprocal of the exponent $\frac{x}{\Delta x}$

That is $\frac{x}{\Delta x} = \frac{\frac{1}{\Delta x}}{x}$

If $\frac{x}{\Delta x}$ is replaced by the symbol u, then $\frac{\Delta x}{x} = \frac{1}{u}$ and the derivative of the natural log of x $= \frac{1}{x} \log \left(1 + \frac{1}{u}\right)^u$

Since $\frac{\Delta x}{x} = \frac{1}{u}$ as x gets closer to zero, u gets bigger and bigger. When x is zero, u is infinitely large.

This complicated-looking expression that results is really simpler than it seems. That is because the expression $\left(1 + \frac{1}{u}\right)^u$ approaches closer and closer to the base of the natural logarithms e as u becomes larger and larger. The value of e can be found by substituting u with bigger and bigger numbers.

When u is 1, then $\left(1 + \frac{1}{1}\right)^1 = (2)^1 = 2$

When u is 2, then $\left(1 + \frac{1}{2}\right)^2 = \left(\frac{3}{2}\right)^2 = 2.25$

When $u = 64$, then $\left(1 + \frac{1}{2}\right)^{64} = \left(\frac{65}{64}\right)^{64} = 2.6973449\ldots$

As u become infinitely large, $\left(1 + \frac{1}{u}\right)^u$ approaches the value of e, which is 2.718281828, and the log of $\left(1 + \frac{1}{u}\right)^u$, being the log of the derived base of natural logarithms e, is simply 1.

The derivative of the log function then turns out to be $\frac{1}{x}\log(e) = \frac{1}{x}$. When the function $y = \frac{1}{x}$ is graphed, e makes another appearance.

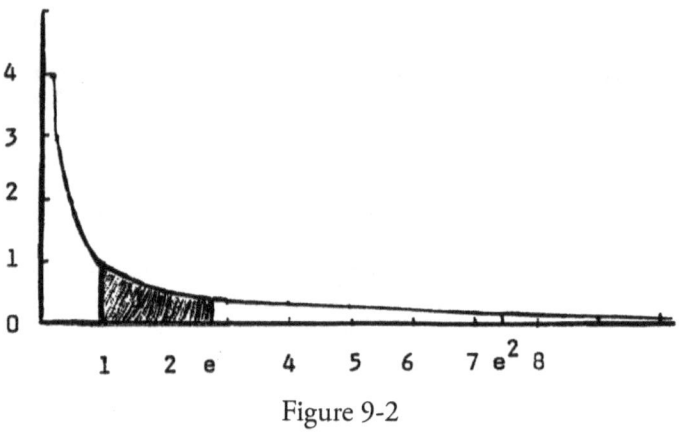

Figure 9-2

The area between the x axis and the curve from $x = 1$ to $x = 2$ is exactly 1. Furthermore, the area between the x axis and curve from $x = e$ to $x = e^2$ is also exactly 1. When the area under the curve from 1 to e is added to itself, the resulting area (2) equals the area under the curve from 1 to e^2. The area under the curve of $y = \frac{1}{x}$ from 1 to any number is the natural log (often abbreviated ln) of the number. It is the exponent that e must be raised to in order to equal the number. So when numbers are multiplied by adding logs, they are really being multiplied by adding areas under the $y = \frac{1}{x}$ curve.

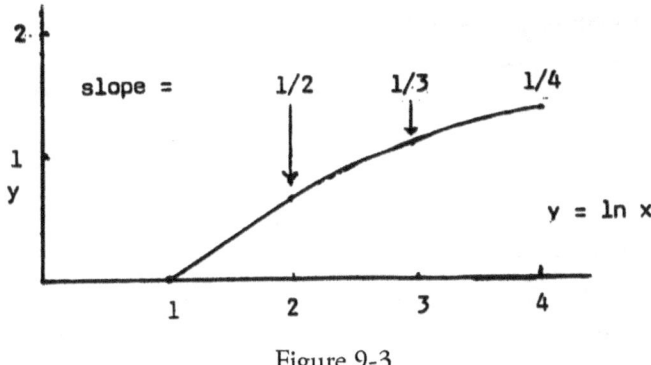

Figure 9-3

The method of finding the value of e from the formula $(1 + x)^{\frac{1}{x}}$ is lengthy and tedious even with a calculator. In the 1600s, finding an accurate value of e using the derivative method was practically absurd. In fact, Newton found the value of e by a different method. Based on ideas borrowed from Nicolaus Mercator, Newton found that the value of e is equal to $1 + \frac{1}{1 \times 1} + \frac{1}{1 \times 2} + \frac{1}{1 \times 2 \times 3} + \frac{1}{1 \times 2 \times 3 \times 4} \cdots$

This calculation is much simpler than the $\left(1 + \frac{1}{x}\right)^x$ method and allows a fairly accurate value for e to be found pretty quickly. The idea for the calculation comes from what is now known as the Maclaurin series. In the case of e raised to a power,

$$e^x = \frac{e^0}{1} x^0 + \frac{e^0}{1} x^1 + \frac{e^0}{(2)} x^2 + \frac{e^0}{(1)(2)(3)} x^3 \cdots$$

$e^0 = 1$ so when $x = 1$, $e^1 = \frac{1}{1} + \frac{1}{(1)(2)} + \frac{1}{(1)(2)(3)} \cdots$

So $e = 1 + 1 + \frac{1}{2} + \frac{1}{6} + \frac{1}{24} + \frac{1}{120} \ldots$ or $e = 1 + 1 + .5 + .167 + .0417 + .0083\ldots = 2.718281828$.

The number e also shows up in situations beyond and seemingly unrelated to logarithms. Leonhard Euler, who gave e its name, showed that $(e^{\pi i}) = -1$. However, discussion of this equation is beyond the scope of this book, principally, because it is beyond the mental capacity of the author.

THE SPEED OF LIGHT

Credit for the first determination of the speed of light is usually given to the Danish astronomer Olaus Roemer, although Roemer's figure was far from today's accepted value. Roemer noticed in 1666 that when Earth, in its orbit, was approaching Jupiter, Jupiter's moons appeared to circle the planet more rapidly than when Earth was receding from Jupiter. When Jupiter is on the far side of the Sun, the orbit of one of its moons, Io, is retarded 16 ½ minutes behind where it appears when Jupiter is on the near side of the Sun. Roemer reasoned that

$$\frac{\text{orbitral period}}{\text{apparent change of orbital period}} = \frac{\text{speed of light}}{\text{speed at which the Earth is moving toward Jupiter}}$$

The method did not give an accurate value for the speed-of-light because the distance to Jupiter used by Roemer was incorrect. He thought delay time to be 22 ½ minutes, as opposed to the correct figure, 16 ½ minutes. It was not until 1761 that the distances to the Sun and planets (and the speed of light) were determined accurately.

A second method of determining the speed of light was discovered in 1725 by English astronomer, James Bradley. Like Roemer's, this method was dependent on knowing Earth's velocity in space. Bradley noticed that stars perpendicular to Earth's orbit seemed to shift 41 seconds of an arc in their position in the sky every 6 months. Bradley reasoned that this stellar aberration, as it is called, resulted from Earth "running into" the starlight in its orbit, so the stars are seen at an angle. The tangent of the angle change observed in 3 months (the time Earth takes to go from zero to maximum perpendicular velocity). It should then be equal to the velocity of Earth divided by the velocity of light.

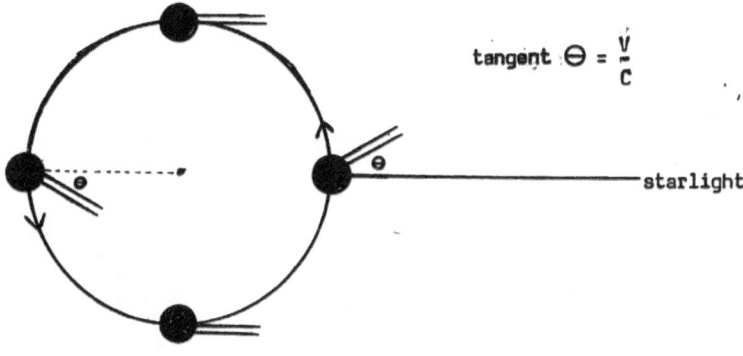

Figure 10-1

The first earthbound determination of the speed of light was made in 1849 by French physicist Hippolyte Fizeau. Fizeau passed a light beam between the teeth of a spinning cogwheel. The light was reflected back from a mirror 5 miles distant. Fizeau determined how fast he had to spin the wheel so that the evenly spaced teeth blocked the light.

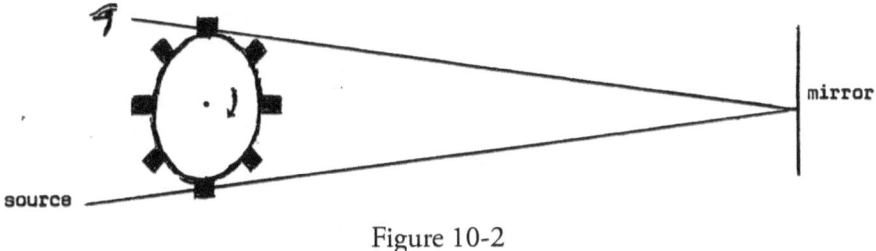

Figure 10-2

In 1850, another Frenchman, Jean Foucault, discovered a superior method. A light source is reflected by a rotating mirror onto a stationary mirror. The stationary mirror reflects the light back to the rotating mirror, from which it is reflected toward an observer at a new angle (due to the rotation of the mirror during the transit of the light). The observer sees the light through a slit. By measuring the speed of rotation at the point at which the observer sees the reflected light (at a particular angle), the time the light traveled between the mirrors is found.

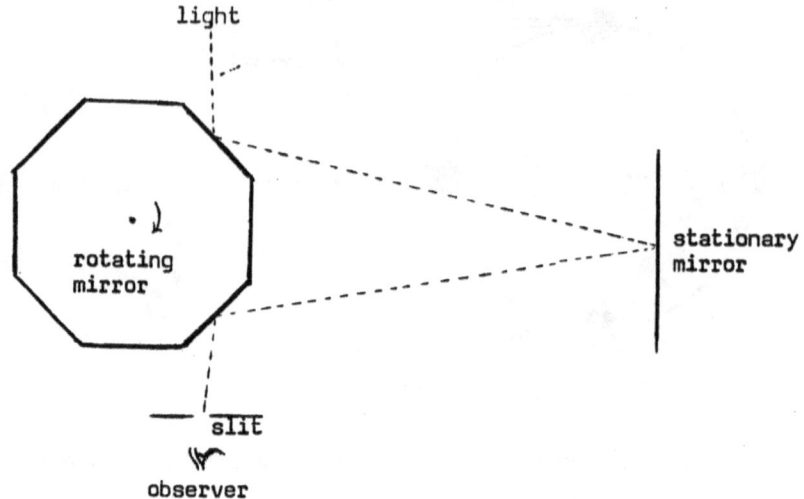

Figure 10-3

DISTANCES TO THE SUN AND PLANETS

Around 250 BC, the Greek astronomer, Aristarchus, attempted to find the relative distances from Earth to the sun and the Moon. He surmised that when the Moon appears exactly half full, the Moon forms a 90-degree angle with Earth and Sun. By measuring the angle of the Moon with respect to the Sun when exactly half of the lunar disc is visible, it should be possible to use the Pythagorean theorem to find the relative distances of the Moon and Sun. Aristarchus measured that angle (incorrectly) as 87 degrees and concluded that the Sun was 19 times as far away as the Moon. Hipparchus later improved on the calculation, but his figure was still far short of the true distance.

After the advent of the telescope, Johannes Kepler noticed that he could detect no parallax (a difference in the apparent position of an object when it is viewed along two different sight lines) of the Sun using the rotation of Earth. If Hipparchus had been right, the parallax would have been detectable. Later, astronomers were able to detect some parallax, but it still wasn't enough to make a good estimate of the Sun's distance.

In 1672, Giovanni Cassini and Jean Richter found the approximate distance to Mars by parallax. Cassini was in Paris and Richter in South America. Knowing the relative distances of the Sun and planets, they obtained a distance of about 85 million miles from the Sun.

In 1690, Edmond Halley proposed a method by which the distance to the Sun could finally be accurately determined. Halley's method was unique in that the distance to the Sun was determined by first finding the diameter of the disc of the Sun. Unfortunately, the method involved the observation of the transit of Venus across the sun, which wasn't to occur until 1761. In 1718, Halley wrote a paper giving the details of the tests to be performed and suggesting that the scientific community make them.

In the Halley method, two widely separated observers on Earth must accurately time the transit of Venus across the face of the Sun. The time will differ for the two observers because, due to their separation, they see Venus crossing the Sun's disc at different solar latitudes. The observer who records a shorter transit time sees Venus transit across a proportionally shorter distance.

Knowing the separation of the observers on Earth gives the distance on the sun's surface that the time difference represents based on transit times. Based on the geometry of a circle, the solar latitude can be calculated based on transit times, the longer transit time corresponding to latitudes closer to the equator. Once the latitudes are calculated, the diameter of the Sun can be found by equating the ratio of the difference of the observed solar latitudes of the Venus transit with the entire diameter of the Sun, which is 180 degrees.

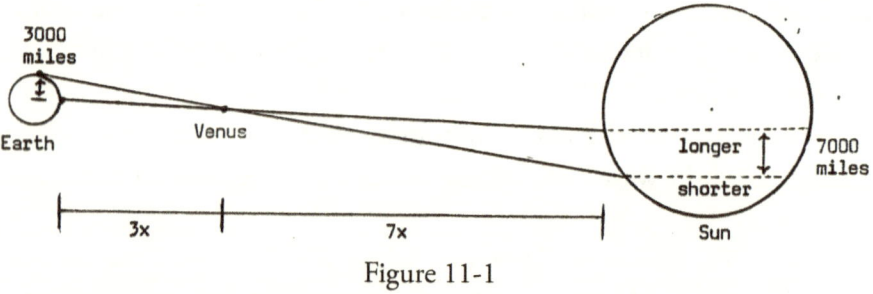

Figure 11-1

Knowing the actual diameter of the Sun and its angular width in the sky, it's a simple matter to find its distance by triangulation.

The sun's diameter is 864,000 miles or 1,390,000 kilometers.

In 1761 and in 1769 (the following Venus transit), a number of transit times were taken at various places on Earth. One was taken on Captain James Cook's first voyage to the South Seas in 1769. A distance of 95 million miles was determined. That number was later revised downward, based on accurate triangulation studies on the asteroid Eros.

Accurate measurements of the distance to the sun can now be made using radar. The distance now agreed upon is an average of 149,597,870,700 meters or 92,955,807 million miles. Exact distance varies due to elliptical orbits.

0 DEGREES, 32 DEGREES, AND 212 DEGREES FAHRENHEIT

There is a popular myth that Gabriel Fahrenheit settled on the zero point for his thermometer on a cold day in Paris. According to the myth, Fahrenheit felt it simply couldn't get any colder, so he calibrated the day's lowest temperature at zero. There is apparently absolutely no truth to this story. In Fahrenheit's time, physicists (natural philosophers) were having great difficulty finding standard calibration points for their newly invented thermometers. The temperature of boiling water, later used as a calibration point by Anders Celsius, wasn't considered usable because the temperature at which water boils varies with respect to altitude and air pressure.

Fahrenheit used three calibration points on his thermometer. For zero degrees, he used the freezing point of a saturated salt solution (salt water so concentrated that no more salt will dissolve). For 32 degrees, he used the freezing point of pure water. Fahrenheit's third calibration point, 96 degrees, was supposed to be the temperature of the human mouth or armpit. Here it is clear that there were still problems in Fahrenheit's day, because using the first two calibration points, our temperatures in the aforementioned places are usually 98.6 degrees. 212 degrees was not an original calibration point on Fahrenheit's scale. It just coincided with the temperature of boiling water.

The thermometer Fahrenheit developed in 1714 became popular (and famous) because it used mercury instead of alcohol. Mercury thermometers have an advantage over alcohol thermometers because mercury's coefficient of thermal expansion (degree of expansion divided by the change in temperature) is more linear than alcohol's. Hence, mercury thermometers are more accurate. Mercury thermometers are also capable of measuring higher temperatures than alcohol thermometers.

THE MASS OF THE EARTH

In the late 1600s, Isaac Newton had determined that the attraction between two masses is equal to a constant multiplied by the product of the two masses (mass is a measure of the amount of matter in an object) divided by the square of the distance between their centers. The trouble was that nobody knew what that constant was. In 1798, Henry Cavendish (who also discovered hydrogen) performed a delicate experiment to determine that constant, which is now known as G. Cavendish suspended a dumbbell on a thin wire. By timing its period of oscillation, he was able to determine how much torque (force) was required to move the dumbbell a certain angular distance. He then placed weights (of known masses), with a known distance from the dumbbell weights (also of known masses), and found how far the attraction between the weights and the dumbbell was able to twist the wire.

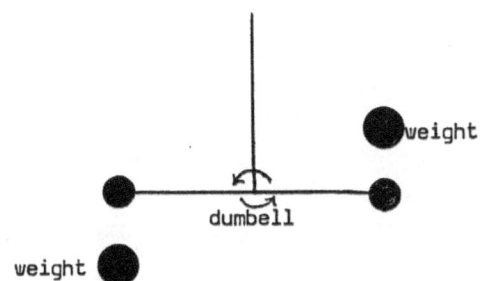

Figure 13-13

He determined G to be $\dfrac{6.67 \times 10^{-8} \text{ cm}^3}{\text{gram second}^2}$

Since force = $\dfrac{(G)(\text{mass 1})(\text{mass 2})}{(\text{distance between the centers of the masses})^2}$ knowing the force Earth exerts on an object of known mass, force = (mass)(acceleration) and acceleration = 9.8 $\frac{\text{meters}}{\text{second}^2}$ and using the radius of Earth as the distance between the two centers of mass, Cavendish found the mass of Earth to be (6×10^{24}) kilograms or (6.6×10^{21}) tons.

THE MASS OF THE SUN

The discovery of the constant G allowed for the immediate calculation of the mass of the sun. The centripetal force on a body in orbit is given by the equation

$$\text{force} = \frac{(\text{mass})(\text{velocity})^2}{\text{distance between the centers of the orbiting and orbited bodies}}$$

Equivalent to Newton's equation, you get

$$\frac{(\text{orbiting body mass})(\text{velocity})^2}{\text{distance between body centers}} = \frac{G\,(\text{mass of sun})(\text{mass of Earth})}{(\text{distance between centers of Earth and sun})^2}$$

From this it is found that Earth's orbital velocity is dependent on the mass of the sun. The equation is…

$$(\text{Earth's velocity})^2 = \frac{G\,(\text{mass of sun})}{\text{distance between Earth and sun}}$$

Since the distance from Earth to the sun had been known since the 1790s (from Halley's method), and since Earth's orbital velocity is known as $\frac{2\pi r}{1\text{ years}}$ plugging in values gives a mass of the sun of about 2×10^{30} kilograms or 2.2 or 10^{27} tons.

MASSES OF PLANETS

The discovery of *G* also allowed the masses of Jupiter, Saturn, and Uranus to be calculated. The distance a moon orbits from a planet can be determined by triangulation using its separation from the planet in space, and the distance to Earth from the planet. The orbital velocity of any moon is also easily found by multiplying the moon's orbital distance by 2π and dividing by the moon's orbital period (found by observation and timing). The planet's mass can then be found in the same way the mass of the Sun was determined by plugging the values into the equation

$$(\text{moon's velocity})^2 = \frac{G \,(\text{mass of planet})}{\text{distance from the moon's center to planet's center}}$$

Mars has two moons, but the determination of its mass by this method had to await their discovery in 1877.

Finding the mass of a moonless planet is more difficult than finding the mass of a planet with moons. In the days before interplanetary rocketry, the only way to guess at a moonless planet's mass was to watch the orbital perturbations caused by the gravitation effects of other planets. Nowadays, we can orbit an artificial satellite around a planet and determine its mass very accurately in the same way the masses of planets with natural satellites (moons) are determined.

ABSOLUTE ZERO

In the seventeenth century, Robert Boyle found that the volume of a gas varies inversely with respect to its pressure. It was also evident that the volume of a gas decreases with decreasing temperature, but the exact relationship was not known. In 1787, Jacques Alexandre César Charles found that the volume of a gas decreases proportionally with decreasing temperature. In 1802, Joseph Louis Gay-Lussac found that to be true of all gases. This is a strange-looking conclusion because it predicts that at some temperature, the volume of a gas will decrease to zero. Of course, gases don't' actually go to zero volume. As they are cooled at some point, they condense to a liquid. But when the volume of any gas is plotted against temperature, the point at which the volume would theoretically go to zero is absolute zero (–459 degrees Fahrenheit or –273 degrees Celsius).

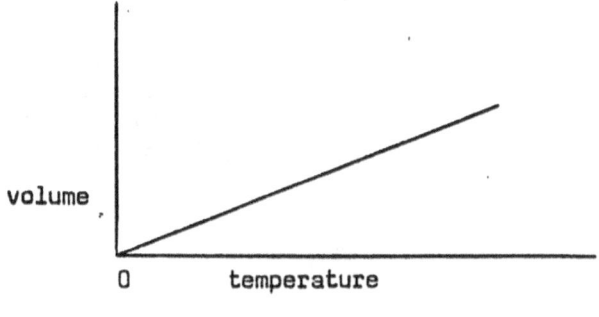

Figure 16-1

DISTANCING TO NEARBY STARS

For centuries, astronomers had attempted to find distances to the stars by finding changes in angles in star positions over a period of six months. This is because in six months' time, Earth is displaced two orbital radii (186 million miles or 300 million kilometers) from its position at the beginning of the period. This is a reasonably long baseline for a triangle. Angles could theoretically be determined by comparing the apparent position of a nearby star with the positions of far more distant stars.

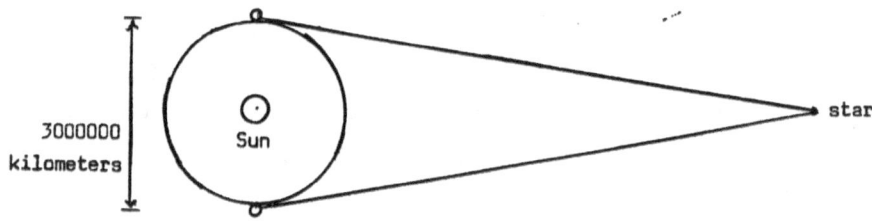

The search for stellar parallax was finally accomplished in 1838 by three astronomers, each of whom calculated the distance to a star. They were German astronomer Freidrich Bessel (61Cygni), Scottish astronomer Thomas Henderson (Alpha Centuri), and Russian astronomer Friedrich Struve (Vega).

ATOMIC AND MOLECULAR WEIGHTS

When John Dalton announced the law of definite proportions and proposed the atomic theory (ca. 1803), he already possessed a list of relative equivalent weights of several substances. These weights were obtained by several methods. In 1774, Antoine Lavoisier introduced a method of obtaining equivalent weights of metals relative to oxygen by weighing a metal, oxidizing it, and then weighing the oxide. For example, if 2.4 grams of a metal are burned and the ash weighs 4.0 grams, then the metal's equivalent weight relative to oxygen is $\frac{2.4 \text{ g}}{4.0 \text{ g} - 2.4 \text{ g}} = 1.5$. Its weight relative to hydrogen (which is 16 times lighter than oxygen) equals 1.5 times 16, which equals 24. About 1790, Claude Louis Berthollet pioneered another method of finding equivalent weights. Berthollet weighed reactant compounds and their products and found the relative weights in much the same way that Lavoisier had with elements. Jeremias Richter extended Berthollet's technique by finding the quantity of a particular base required to completely neutralize a known quantity of a particular acid. Richter's value for the equivalent weight of sulfuric acid was 1000. Consequently, if it required two grams of a base to completely neutralize one gram of sulfuric acid, his equivalent weight for the base would be 2000. Using this technique, Richter found the equivalent weights of eight bases and thirteen acids.

In Dalton's time, the concepts of equivalent, atomic, and molecular weights were confused and intermingled.

The first real verification of Dalton's atomic hypothesis came about as a result of an 1818 physics experiment performed by Pierre Dulong and Alexis Petit. One of the physical characteristics of a substance is called specific heat, which is the amount of energy (measured in calories) required to raise the temperature of one gram of a substance one degree Celsius. It can be measured by placing a weighed quantity of the substance in a measured quantity of water at a different temperature and observing the change in water temperature. The higher the substance's specific heat, the more energy it will draw from the water, and the greater the water temperature change will be. Dulong and Petit found that the higher a metal's atomic weight, the lower its specific heat according to the equation atomic weight = $\frac{6.3}{\text{specific heat}}$

As an example, suppose twenty grams of a metal at 95 degrees C is added to fifty grams of water at 20.0 degrees C and the water is heated to a final temperature of 23.1 degrees C, if the water container is insulated and has a low specific heat itself, then the energy gained by the water should be very close to the energy lost by the metal. The energy gained by the water is $\left(\frac{1 \text{ calorie}}{\text{gram degree C}}\right)$ (50 grams)(3.1 degrees C) = 155 calories. Setting this equal to the energy lost by the metal, we have 155 calories = (20 grams) (95 degrees C - 23.1 degrees C) (the metal's specific heat)—155 = (20) (71.9) (specific heat)—so the metal's specific heat = .108. Plugging into Dulong and Petit's formula, atomic weight = $\frac{6.3}{.108}$ = 58, the metal is, therefore, probably either nickel (atomic weight 58.71 g/mole) or cobalt (atomic weight 58.93 g/mole).

By the 1820s, the atomic theory had caught on, and in 1826, Jöns Jacob Berzelius published an accurate chart of atomic weights listing 47 elements. Berzelius based his atomic weights on oxygen having an atomic weight of exactly 100. Oxygen has a weight of 15.999 on our charts, so although Berzelius chart was accurate, it was quite different form the charts we see today.

The Berzelius system persisted until 1858 when Stanislao Cannizzaro revolutionized chemistry (and physics), realizing that Avogadro's assumption that equal volumes of gasses contain equal numbers of particles was true. Cannizarro based his scheme of atomic masses on the mass of the lightest gas—hydrogen. On the basis of Cannizzaro's findings, the mole was defined as the number of hydrogen atoms contained in one gram of hydrogen. Alternatively, the mole was defined as the number of molecules of any gas contained in 22.4 liters at one

atmosphere pressure and at 0 degrees C (273.15 on the Kelvin temperature scale). The atomic masses we see in reference books today are very close to Cannizzaro's. With Canizzarro's discovery, gas analysis became a powerful analytic tool. Formula and molecular masses of compounds that react to form gases could be found by weighing samples and measuring gas volumes or by weighing residues and measuring gas volumes. For instance, under certain conditions, ammonium compounds react with hypochlorite ion to form nitrogen gas, according to the equation:
$2NH_4^+ + 3OCl^- \rightarrow N_2\uparrow + 3Cl^- + 3H_2O + 2H^+$.

(Caution: Do not try this. Under other conditions, reactants may form poisonous chloramine gas or explosive nitrogen trichloride.) If 2.0 grams of ammonium chloride are reacted in excess sodium hypochlorite solution, 419 milliliters of N2 (nitrogen gas) at one atmosphere pressure and degrees C is produced. Using this information, the formula mass of ammonium chloride may be found:

$$\left(\frac{1 \text{ mole N2}}{2 \text{ moles NH4}^+}\right)\left(\frac{22400 \text{ ml N2}}{1 \text{ mole N2}}\right)\left(\frac{2.0 \text{ grams salt}}{419 \text{ ml N2}}\right) = \frac{53.5 \text{ grams salt}}{\text{mole NH4}^+}$$

This corresponds to the formula of NH_4Cl.

As good as the techniques had become in Cannizzaro's time, there were still classes of compounds (mainly organic) that defied all attempts to discover their molecular weights. This problem was solved in 1882 when François-Marie Raoult observed that the freezing point of a solvent decreases in a regular way as more solute is dissolved in it. For example, the freezing point of a kilogram of water decreases 1.86 degrees C for every mole of particles dissolved. If 120 grams of a molecular compound dissolved in a kilogram of water lowers its freezing point to −3.72 degrees C, then the formula mass of the compound is 60 grams/mole.

$$(3.72 \text{ degrees C}) \left(\frac{1 \text{ mole of particles}}{1.86 \text{ degrees C}}\right) = 2 \text{ moles of particles}$$

120 grams/2 moles = 60 grams/mole

Different solvents have different constants (1.86 degrees C/mole kilogram is water's constant) but the method for finding the molecular weight of a compound is the same for any solvent. Raoult's method was the principal technique used to find molecular weights until mass spectrometry was developed in 1919 by Francis W. Aston. In the mass spectrometer, the substance being analyzed is ionized, and its positively charged constituent particles passed between charged plates. The higher the charge-to-mass ratio a particle has, the greater deflection it will undergo. Atomic and molecular masses can be easily determined to one part in 10,000 with a mass spectrometer.

THE GAS CONSTANT R

In the early 1800s, Joseph Louis Gay-Lussac discovered that gases unite with each other in proportion to their volumes, and that the volumes of the resultant gases are also proportional to the volumes of the reactant gases. For example, at constant temperature and pressure, one liter of hydrogen gas combined with one liter of chlorine gas to produce two liters of hydrogen, chloride is $Cl_2 + H_2 = 2HCl$. The volumes are not always additive, however, three liters of hydrogen gas combined with one liter of nitrogen gas to produce two liters of ammonia gas is $3H_2 + N_2 = 2NH_3$.

Having read Gay-Lussac's paper on gas volumes, Amadeo Avogadro proposed in 1811 that equal numbers of particles of different gases occupy equal volumes at the same given temperatures and pressures. Avogadro's hypothesis was generally ignored by his contemporaries, and it was not until 1858 that Stanislao Cannizzaro firmly established the theory of gases. Using Boyle's law (initial gas pressure x initial gas volume = final gas pressure x final gas volume) and Charles's law (initial gas volume/final gas volume = initial gas temperature/final gas temperature), Cannizzaro derived the combined gas law:

$$\text{final gas volume} = (\text{initial volume}) \frac{\text{initial pressure}}{\text{final pressure}} \frac{\text{final temperature}}{\text{initial temperature}}$$

Using Avogadro's hypothesis with the combined gas law, Cannizzaro also refined the ideas of atomic and molecular weights, basing them on multiples of the atomic weight of hydrogen. On the basis that a gram of hydrogen contains what is now referred to as a mole of hydrogen atoms, Cannizzaro discovered that a mole of any kind of gas particles always occupies close to 22.4 liters at one atmosphere of pressure (760 mm) and 0 degrees Celsius (32 degrees Fahrenheit). Cannizzaro's atomic weights are very close to the atomic weights we use today. After Cannizzaro's discovery, it rapidly became apparent to physical chemists that when the fact that a mole of gas occupies 22.4 liters at one atmosphere and 0 degrees C (273 K) is incorporated into the combined gas law, a more useful equation results. It is derived as follows: Taking Boyle's law, pressure x volume = a constant (at constant temperature), and Charles's law, volume = a constant x temperature (at constant pressure), an equation of the form pressure x volume = a constant x temperature (PV = KT) results.

At constant temperature and pressure, the volume of a gas depends only on the number of moles of gas particles (i.e., volume = a constant x moles). Combining this fact with the above equation, pressure x volume = a new constant x moles x temperature. The resulting equation (PV = nRT) is called "the ideal gas law," and the new constant, abbreviated R is the gas constant (.082 liters atmospheres/mole K).

Rudolph Clausius is officially credited with the creation of the ideal gas law, although the development of the law took several years, and other chemists were involved. Soon after the development of the law, however, problems appeared. Many gases didn't fit the equation exactly. In 1873, Johannes van der Waals modified the equation so that it would work for all gases. The first part of the problems was the size of the gas molecules themselves (ideal gases are made of infinitely small particles). And the other part of the problem was the gas molecules' mutual attraction. The final form of the equation was (pressure + $\frac{\text{an inter-molecular attraction constant}}{(\text{volume of gas})^2}$) (gas volume - volume of molecules) = moles (R) (T)

R turns out to be a special constant because it represents the amount of energy a mole of gas particles absorbs with a rise in temperature of 1 degree C. This fact involves R in a number of thermodynamic equations concerned with energy transitions. In chemistry, for example, it is possible to tell whether a reaction will take place spontaneously or not according to the equation.

Free energy = – R(temperature) (the natural log of the equilibrium constant)

If the free energy is negative, the reaction will go. Even more important than this equation is the fact that R appears in the equation that led Max Planck to the quantum theory. So in a very real sense, we owe the development of much of modern physics to the discovery of the ideal gas law.

The accepted value for R is 8.31441 joule/mole K or .082057 liter atmospheres/mole K.

THE MASS OF THE MOON

Because the Moon has no natural satellite of its own, the determination of its mass was a difficult problem. The Sun's mass, the planets' masses, and even the masses of some binary stars were known before anyone knew the mass of the Moon.

In the 1700s, James Bradley discovered that the stars seemed to shift position in the sky several arc seconds every half month. He attributed the apparent shift to an effect of the Moon. The American astronomer, Simon Newcomb, compiled accurate enough data by observing the apparent shift in the position of the Sun to calculate the mass of the Moon.

Treating the Earth and Moon as a two-body system, as the Earth orbits the Sun, the Earth is also orbiting around the center of mass of the Earth-Moon system. The Earth and the Moon both have mass, so instead of the Moon orbiting around the center of the Earth, in a sense, they orbit around each other. The Earth is much more massive than the Moon, so it appears (without careful measurements) that the moon has no effect on the Earth. But in fact, the Moon and Earth revolve around a center of mass, which is not the center of the Earth but is within the interior of the Earth.

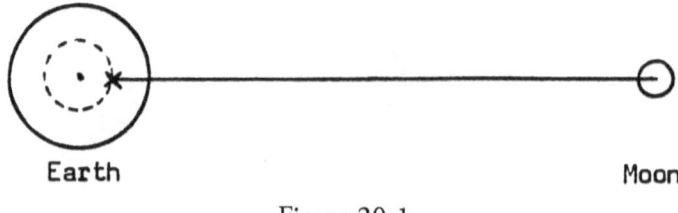

Figure 20-1

It is the center of the Earth-Moon system, rather than the center of the Earth itself, that describes an elliptical orbit around the sun. The Earth and Moon together revolve about their common center of mass. This point lies inside Earth about 4,800 km (3,000 miles) from its center. It is this common center of mass, not the center of Earth, that follows an elliptical path around the sun.

The Earth-Moon system may now be treated as if it were on a balance. The mass of Earth times the distance to the center of Earth-Moon mass (2890 miles) equals the mass of the moon times its distance to the center of Earth-Moon mass (235970 miles).

$$(5.983 \times 10^{24} \text{ kg})(2890) = (235970 \text{ miles})(\text{mass of moon}).$$

$$\text{Mass of moon} = 7.33 \times 10^{22} \text{ kg. } (2{,}890 \text{ miles})$$
$$E = 238{,}860 \text{ miles} - 2{,}890 \text{ miles}) (M)$$

$$(2{,}890) (5.983 \times 10^{24} \text{ kg}) = (23{,}590) (M). \quad M = 7.33 \times 10^{22} \text{ kilograms}$$

We can now, of course, make an artificial satellite orbit the moon and, from its velocity and distance, find the mass of the moon more accurately. The accepted mass of the Moon is $7.34767309 \times 10^{22}$ kilograms.

THE RYDBERG CONSTANT

In 1859, Gustav Robert Kirchhoff proposed that it might be possible to identify elements in the Sun's atmosphere by comparing its spectrum with the spectra of known gases. In 1868, G. Johnstone Stoney noticed there was a kind of pattern in the spectrum of hydrogen. In 1885, that pattern was discovered and reported by Johann Jakob Balmer, a school teacher in Basel, Switzerland. Balmer had inherited an interest in numbers from his great uncle, who dabbled in numerology. One day while talking to a physicist friend, Balmer complained of boredom. Aware of Balmer's interest in numbers, the friend suggested that he tackle the problem of the hydrogen spectrum, and gave him the wavelengths of some of the lines. Balmer noticed that when the wavelengths were expressed as a fraction multiplied by a constant, the numerators were squares of integers, while the denominators were always four less than the numerators.

For example, the wavelengths $\frac{9}{5}$ K, $\frac{16}{12}$ K, and $\frac{36}{32}$ K can be expressed as $\frac{3^2}{3^2-2^2}$ K, $\frac{4^2}{4^2-2^2}$ K, $\frac{5^2}{35-2^2}$ K, $\frac{6}{6^2-2^2}$ K...

But new discoveries were being made that didn't fit into Balmer's formula, and it became apparent in 1888 that the Balmer series was a special case.

In 1890, Swedish physicist, Johannes Rydberg, reported that he was using a variation of Balmer's formula in which all the lines of the hydrogen spectrum apparently fit. Rydberg's formula was

$$\frac{1}{\text{wavelength}} = \text{a constant} \left(\frac{1}{(\text{an integer})^2} - \frac{1}{(\text{another integer})^2} \right)$$

*The constant is known as Rydberg's constant.

Later, Theodore Lyman IV discovered lines in the ultraviolet region that didn't fit the Balmer formula but were predicted by Rydberg's. Friedrich Paschen and F. S. Brackett found additional lines in the infrared region that Rydberg's formula had predicted. In fact, Rydberg's formula accurately predicts all the lines in the hydrogen spectrum.

Lyman	$\frac{1}{\text{wavelength}} = R\left(\frac{1}{1^2} - \frac{1}{n^2}\right)$ n may be 2, 3, 4, etc.
Balmer	$\frac{1}{\text{wavelength}} = R\left(\frac{1}{2^2} - \frac{1}{n^2}\right)$ n may be 3, 4, 5, etc.
Paschen	$\frac{1}{\text{wavelength}} = R\left(\frac{1}{3^2} - \frac{1}{n^2}\right)$ n may be 4, 5, 6, etc.
Brackett	$\frac{1}{\text{wavelength}} = R\left(\frac{1}{4^2} - \frac{1}{n^2}\right)$ n may be 5, 6, 7, etc.

Later, similar formulas involving integers and differences were found to predict spectral lines of other elements.

The significance of Rydberg's discovery was not realized at the time. Niels Bohr finally explained in 1913 that the lines in the hydrogen spectrum resulted from allowable quantized jumps of electrons from higher suborbitals to lower ones, and that the spacing between the lines represented the areas of forbidden transitions.

Louis de Broglie showed in 1924 that the only possible electron suborbitals are integral (whole) numbers of wavelengths in circumference.

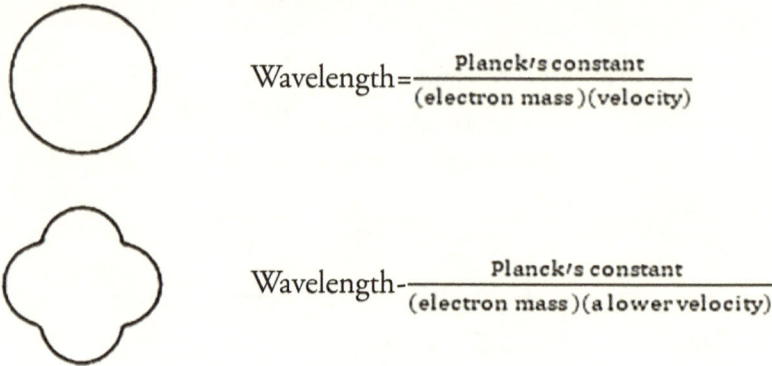

Figure 21-1

The lines of the spectra result from photons of light given off from an atom when the orbit of an excited electron decreases in size from some (whole) number of wavelengths to a lesser number.

PLANCK'S CONSTANT, THE STEFAN-BOLTZMANN CONSTANT, WIEN'S DISPLACEMENT CONSTANT, AND BOLTZMANN'S CONSTANT

Planck's constant, usually abbreviated as h, is found in literally every phase of the science of the very small. The energy of a particle of light is given by the equation: energy = (Planck's constant) (frequency of light).

The circumference of an unexcited hydrogen atom can be found from the equation

$$\text{circumference} = \frac{\text{Plank's constant}}{(\text{electron mass})(\text{electron velocity})}$$

The constant is found in equations giving specific heats of metals, equations that tell us the limitations on the measurements we can make, equations that tell what intensities and frequencies of light are given off by a glowing object, and many more.

Considering all the equations in which Planck's constant plays a role, it should have been a simple matter to find its value accurately. However, Max Planck found his constant the hard way. Not only was the physics that led to the discovery difficult, it was also confusing.

We have all seen the glow emitted from the coils of an electric stove or from metal that is being welded. The glow that hot objects give off is called black-body radiation and was named by Gustav Robert Kirchhoff in a paper written in 1860. In 1879, after having read about John Tyndall's experiments with hot platinum wire, Josef Stefan noticed that the intensity of the total black-body radiation from a glowing object is proportional to the object's absolute temperature to the fourth power.

Tyndall had reported that the radiation intensity of platinum at 1,473 K was 11.7 times the intensity at 798 K. Stefan recognized that $(\frac{1,473}{798})^4$ is about 11.7, so his equation says that intensity = (a constant) (temperature)^4. The constant, is known as the Stefan-Boltzmann constant (not to be confused with Boltzmann's constant). It is represented by the symbol a and has a value of $5.67051(19) \times 10^{-5}$ erg cm^{-2} K^{-4} s^{-1}.

Stefan's discovery stimulated interest in finding radiation intensities at each wavelength of the light emitted.

In 1879, radiation measurements were made with a crude device called a thermopile, but the following year, Samuel Pierpont Langley invented a much more sensitive device, which he called a bolometer. This invention greatly facilitated research on black-body radiation intensities.

In 1893, Wilhelm Wien discovered that the wavelength of maximum intensity times the absolute temperature of a glowing body equals a constant. This constant b is known as Wien's displacement constant and is equal to 2.8978×10^{-3} meters K. Despite Wien's success, he was having difficulty explaining the curve of the distribution of radiation intensities with respect to frequencies. V. A. Michelson thought he knew a way the problem could be approached theoretically. Michelson thought (incorrectly) that if black-body radiation resulted from the vibrations of atoms, then James Clerk Maxwell's formula for the velocity distribution of particles of a gas would hold for a black-body ratio as well. Following Michelson's lead, Wien came up with the equation. The intensity of radiation at a particular wavelength equals

$$\frac{\text{a constant}}{(\text{wavelength})^5} \; e^{\frac{\text{another constant}}{(\text{wavelength})(\text{absolute temperature})}}$$

(Note: In the above equation, e is the base of the natural logarithms.)

This equation, known as Wien's law, was later amended by Planck.

Experiments testing short wavelengths of light indicated Wien's law worked despite the fact that (as was later discovered) the theoretical basis for it was wrong, Wien's law met with difficulties when experimentalists, Otto Lummer and Ernst Pringsheim Sr., found that it didn't hold for longer wavelengths. About the same time, British mathematician, Lord Rayleigh found, that Wien's equation had theoretical as well as experimental problems. But when Rayleigh derived what should have been the correct equation, energy density = $\frac{8\pi \, (\text{a constant}) \, (\text{temperature})}{(\text{wavelength})^4}$ it didn't work for short wavelengths. In fact, it predicted that at short wavelengths, the energy density should become infinitely large. Strangely, Rayleigh's equation (corrected by James Jeans) did accurately predict energy intensities at lower frequencies. Planck realized that it was necessary to reach some kind of a compromise between the two equations since one appeared to fit reality at the high end of the spectrum and the other at the low end. In October 1900, Planck derived an equation that gave the Rayleigh-Jeans curve at the low end of the spectrum and the Wiens curve at the high end: intensity = $\frac{\text{a constant}}{(\text{wavelength})^5 \, \left(e^{\frac{\text{a constant}}{\text{wavelength} \times \text{temperature}}} - 1\right)}$

(Note: In the above equation, e is the base of the natural logarithms.)

In an attempt to find out why his formula worked, Planck began looking at entropy, the tendency of all isolated process to become random. Planck was an expert in this area. In fact, it was the results of his previous calculations on entropy that led him to the "compromise" equation. In looking at Boltzmann's equation, (entropy = a constant x the log of the number of energy distributions), Planck noticed that there had to be a finite number of energy elements for the equation to work, because if the number of energy distributions were infinite, the entropy would also be. This led Planck to the conclusion that a particle cannot transfer less than a certain amount of energy to another particle, and that when energy is transferred from one particle to another, it is transferred in whole multiples of that minimum energy (Planck's constant x particle oscillation frequency).

In its final form, Planck's equation (based on his assumption of quantized energy) introduces two new constants. Radiation energy at a particular wavelength equals

$$\frac{(2\pi)(h)(\text{the speed of light})^2}{(\text{wavelength})^5 \; \dfrac{e^{(h)(\text{speed of light})}}{[(k)(\text{wavelength})(\text{temperature})] - 1}}$$

In the above equation, h is Planck's constant (6.62176×10^{-34}) joule-second, and k is Boltzmann's constant, which is the energy an atom gains with an increase of temperature of 1 degree C. The value of k is (1.380662×10^{-23}) joules/K.

THE SURFACE TEMPERATURE OF THE SUN

When light emitted from hot, glowing gases is passed through a prism or diffraction grating, line spectra (bright lines of specific frequencies)—characteristic of the elements present—are observed. Each element has its own unique spectrum. Being gaseous, the Sun also produces line spectra. By comparing the Sun's spectrum with spectra made by hot, glowing gases on Earth, it is possible to tell what elements are present in its atmosphere.

Because the Sun is opaque, it produces a continuous black-body spectrum in addition to the line spectra. Consequently, when Josef Stefan discovered that the total radiance of a black-body radiator is proportional to its temperature to the fourth power (i.e., a constant x (temperature)^4), he also discovered a way to find the temperature of the Sun's surface. When Wilhelm Wien discovered in 1893 that the temperature of a black-body radiator multiplied by the wavelength of maximum intensity equals a constant (Wien's displacement law), he found another (more accurate) way to measure the sun's surface temperature. In 1902, using Stefan's method, the Sun's surface temperature was measured at 5770 degrees C. In 1903, based on experimental findings that the wavelength of maximum intensity in the Sun's radiation is 4,900 angstroms (yellow-green), Samuel Pierpont Langley and George Abbot found the Sun's temperature to be 5,920 degrees C.

- (Wavelength)(temperature) = Wien's constant (b)
- b = 2.8978×10^{-3}
- (4.9×10^{-7} meters)(temperature) = 2.8978×10^{-3}
- Temperature = 5900 K

Today the sun's surface temperature is estimated to be about 5,800 K.

AVOGADRO'S NUMBER

Lorenzo Romano Amadeo Carlo Avogadro never knew his number, nor did anyone else, until almost five decades after his death. In 1811, Avogadro proposed that equal volumes of gases contain the same number of particles when at the same temperature and pressure, but he did not know what the number was for a given volume.

The mole, the quantity of material that contains Avogadro's number of particles, was originally defined as the number of hydrogen atoms in exactly one gram of hydrogen. The fact that hydrogen has three isotopes resulted in problems when the mole was defined in this way, so the definition of the mole has since been slightly modified. Avogadro's number is now defined on the basis of the carbon atom. A carbon atom, possessing six neutrons, has an atomic mass of EXACTLY twelve.

The discovery of Avogadro's number stands out as a scientific anomaly in that decades before it was found, all the information was available to get at least a rough approximation of it. Benjamin Franklin had discovered that certain oils will spread out over the surface of water to form an extremely thin film. For oils such as oleic acid, the film is exactly one molecule thick. The oil usually spreads out in a generally circular fashion, so the thickness of a monolayer of it can be found using the formula for the volume of a cylinder: volume of oil (known) = π (radius of layer)^2(thickness).

Lord Rayleigh and Wilhelm Roentgen used this method to estimate the size of a molecule in 1890 but, strangely, never used it to calculate the number of atoms in a mole. In fact, Avogadro's number was never calculated in the nineteenth century despite the possibility of doing so with any of several methods.

The first estimate of the size of a molecule (which would have made the calculation of Avogadro's number possible) was made in 1865 by Josef Loschmidt. Working from equations derived by James Clerk Maxwell and Rudolf Clausius, Loschmidt calculated that the diameter of a gas molecule is equal to 8 (a condensation coefficient—the mean free path of the molecule). The condensation coefficient is roughly the ratio of the volume a liquid occupies at standard temperature and pressure, and the volume of its gas under the same conditions. If he had used his calculations to attempt to calculate Avogadro's number, Loschmidt would have arrived at the conclusion that a mole was equal to 4.1×10^{23} particles.

By 1870, Lord Kelvin had found three other methods that gave estimates of the sizes of molecules, but none of them was very accurate. They came from

1. light scattering considerations;
2. electrical attractions between metal plates; and
3. the idea that soap bubbles are one molecule thick.

Again, no effort was made to go beyond determining the sizes of molecules.

Finally, in 1900, Max Planck calculated a theoretical value for Avogadro's number. In the final form of Planck's black-body equation, Boltzmann's constant k (the energy increase of a single molecule of gas with a rise of temperature of 1 degree Celsius) was the only unknown. The value is about 1.38×10^{-23} joules/K. From the ideal gas equation (pressure)(volume) = (the gas constant R)(moles)(temperature), Planck knew that the amount that the energy of a mole of gas molecules increases with a temperature rise of 1 degree Celsius = R or 8.3 joules/K. Planck calculated that the number of molecules in a mole is

$$\frac{8.3 \frac{\text{joules}}{\text{mole K}}}{1.38 \times 10^{-23} \frac{\text{joules}}{\text{molecule K}}} = 6.0 \times 10^{23} \frac{\text{molecules}}{\text{mole}}$$

Planck's calculation, however, was not taken seriously at the time by most scientists. In fact, in 1900, some famous scientists (e.g., Wilhelm Ostwald and Ernst Mach) didn't even believe in the atomic theory. Albert Einstein's calculation of Avogadro's number in 1903 apparently still didn't convince everyone.

The first accepted value of Avogadro's number resulted from an experiment made in 1908 by Jean Perrin. In 1905, Albert Einstein derived an equation that linked Avogadro s number with the diffusion rate of small particles in a liquid. In Perrin's work based on that equation, Avogadro's number equals

$$\frac{(\text{the gas constant R})(\text{the absolute temperature})}{(6\pi)(\text{particle diffusion rate})(\text{ratios of particles})(\text{viscosity of liquid})}$$

Perrin's experiments with diffusion rates of resin particles in water gave him a figure of 6.85×10^{23} particles/mole.

Avogadro's number was finally determined accurately in 1910, when Robert Millikan found the charge of an electron. In the early 1800s, Michael Faraday found that it required 96,494 coulombs of electricity to electroplate one mole of silver atoms. When Millikan found that charge on a single electron is 1.6×10^{-19} coulombs, it became a simple matter to calculate Avogadro's number:

$$\frac{1 \text{ electron}}{1.6 \times 10^{-19} \text{coulombs}} \frac{96494 \text{ coulombs}}{\text{mole}} = \frac{6.03 \times 10^{23} \text{ electrons}}{\text{mole}}$$

More accurate measurements of electron charges have refined Avogadro's number downward slightly from the older values. Today's value is $6.02214129 \times 10^{23}$.

CHARGE AND MASS OF AN ELECTRON

In 1897, Joseph John "JJ" Thomson was able to measure the ratio of an electron's charge e to its mass m by deflecting a beam of electrons with electric and magnetic fields. The deflection of the beam was detected by allowing it to strike a fluorescent screen. Thomson did it the hard way. The method used for finding e/m today is simpler than Thomson's method simply because it looks at the curvature of a beam of electrons in a magnetic, rather than an electric, field. This method results in simpler, more straightforward equations. Both methods will be described here.

In today's method, when a beam of electrons is bent with a magnetic field, the curved portion can be compared to a segment of a circle with a radius of r.

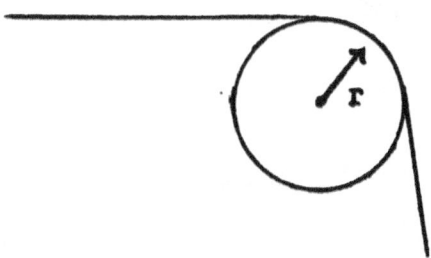

Figure 25-1

The equation that relates this curvature to elementary charge and electron mass is

$$\frac{e}{m} = \frac{\text{elementary charge}}{\text{electron mass}} = \frac{\text{velocity of electrons}}{(\text{strength of magnetic field})(\text{radius of curvature})}$$

The only unknown quantity on the right side of the equation is the velocity of the electrons.

Besides being deflected by a magnetic field, electrons can also be deflected by an electric field. When an electron passes between two oppositely charged plates, it feels a force given by the equation

$$\text{Force} = \frac{(\text{elementary charge})(\text{voltage on plates})}{\text{distance between plates}}$$

In a magnetic field, the force an electron feels is given by the equation:

Force - (elementary charge)(electron velocity)(magnetic field strength)

If magnetic and electric fields are set up so that they force the electrons in exactly opposite directions, and the electric field is adjusted by changing the voltage on the plates until the electron beam is exactly straightened out, then the magnetic force equals the electric force, and

$$\frac{(\text{elementary charge})(\text{voltage on plates})}{\text{distance between plates}} = (\text{elementary charge})(\text{electron velocity})(\text{magnetic field strength})$$

This reduces to the equation:

$$\text{velocity of electrons} = \frac{\text{voltage on plates}}{(\text{field strength})(\text{distance between plates})}$$

Once the electron velocity is found in this manner, it can be plugged into the *e/m* equation. Thomson obtained a value for e/m of 1.7 times 10^{11} coulombs/kilogram.

In Thomson's method, when J. J. Thomson first performed the *e/m* experiment in 1897, he observed the deflection of an electron beam due to an electric field (rather than magnetic). The equation for this is

$$\text{deflection} = \frac{(\text{elementary charge})(\text{charge field strength})(\text{length of plates})^2}{2(\text{mass of electrons})(\text{velocity of electrons})^2}$$

Again, the only unknown quantity is the velocity of the electrons. In the original experiment, Thomson applied the electric and magnetic fields one at a time. He found the electron velocity by adjusting the magnetic field until it gave the same beam deflection that the electric field had produced. At that point, the electrons felt the same force from the magnetic field as they had felt in the electric field, and the following equation could be used:

$$\text{velocity of electrons} = \frac{\text{voltage on plates}}{(\text{magnetic field strength})(\text{distance between plates})}$$

Thomson's data gave elementary charge/electron mass, but it was necessary to know either the elementary charge was to find the mass of the electron or to know the mass of the electron to find the elementary charge.

Max Planck's theoretical value for Avogadro's number (from his black-body equation) and the gas constant R had given him a handle on the elementary charge e. Planck knew that Faraday had found that it required about 96,500 coulombs of electricity to plate out a mole of a univalent metal. Now that Planck thought he knew the number of particles in a mole, he was able to find the charge on an electron theoretically.

$$\left(\frac{1 \text{ mole}}{6.0 \times 10^{23} \text{ electrons}}\right)\left(\frac{96500 \text{ coulombs}}{\text{mole}}\right) = 1.6 \times 10^{-19} \frac{\text{coulombs}}{\text{electron}}$$

However, since Planck's numbers were thought to be based on shaky theory, most scientists waited for experimental verification of Planck's value for the elementary charge before they would accept it. This delayed the official recognition of the values of e and m for 10 years.

Experimental evidence for the absolute values for both, the elementary charge and the electron mass, was finally determined in 1910 by an experiment performed by American physicist, Robert Millikan.

Recall from the Thomson experiment that the force on an electron in an electric field equals the elementary charge multiplied by the voltage across the plates divided by the distance between the plates. Millikan induced charges on tiny oil droplets by exposing them to X-rays. He then found the charges on them by just stopping the fall of some of the droplets, balancing the gravitational force on them with an opposing electrical force. At the point where the electric force and gravitational force are equal,

$$\frac{(\text{charge on droplet})(\text{voltage across plates})}{\text{distance between plates}} = (\text{mass of droplet})(\text{gravitational acceleration})$$

So

$$\text{charge on droplet} = \frac{(\text{mass of droplet})\left(9.8 \, \frac{\text{meters}}{\text{second}^2}\right)(\text{distance between plates})}{\text{voltage across plates}}$$

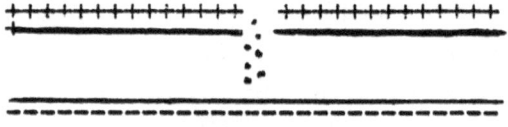

Figure 25-2

Millikan was able to find the mass of the oil droplets from the density of the oil and the radii of the droplets (found from their terminal velocities in air). Some of the droplets were singly charged, so Millikan could measure *e* directly. Other droplets were multiply charged. The fact that there were no fractions of charges showed that electric charge is quantized.

The modern value for the electron mass is $9.10938291 \pm 0.00000040 \times 10^{-31}$ kg.

The modern value for an elementary charge is $1.602176565 \times 10^{-19}$ coulomb.

THE SIZE OF ATOMS

The discovery of accurate sizes of atoms cannot rightfully be assigned to any particular person. In fact, there is some ambiguity as to what actually constitutes atomic radius. The electron density of an atom does not cut off sharply at any particular point.

Josef Loschmidt's work in 1865 gave the first rough estimates of the sizes of molecules (see the chapter on Avogadro's number). However, accurate determinations of atomic dimensions had to await further experimental developments.

Early in the twentieth century, English physicists Henry and Lawrence Bragg conducted x-ray diffraction studies to find atomic radii. About the same time, other people were finding the radii of atoms by calculation based on their interatomic spacings. This became possible, of course, as soon as an accurate version of Avogadro's number became available.

One can get a rough approximation of interatomic spacings by treating atoms as if they were cubes rather than spheres. Knowing, for example, that the density of aluminum is 2.702 grams/cc, the approximate number of atoms in one cc can be calculated as $\left(\frac{2.7 \text{g Al}}{1 \text{ cc Al}}\right)\left(\frac{1 \text{ mole Al}}{27 \text{g Al}}\right)\left(\frac{6.0 \times 10^{23} \text{ atoms}}{1 \text{ mole Al}}\right) = 6 \times 10^{22} \frac{\text{atoms}}{\text{cc}}$

The volume of a single atom of aluminum then is $\frac{1 \text{ cc}}{6 \times 10^{22}} = 16.7 \times 10^{-24} \text{cm}^3$

The diameter of an aluminum atom is approximately the cube root of the volume, which is about 2.6×10^{-8}cm.

Atoms, however, are spheres rather than cubes, so the atomic radius calculated above is not quite right, in order to find interatomic distances accurately, it is necessary to know the crystalline structure of the material and to invoke a concept called the unit cell. Materials made of crystals of atoms packed in the form of a cube, such as aluminum, may have one of two structures. These are known as body-centered cubic, where a central atom is enclosed in a cubic "cage" made out of 8 surrounding atoms, and face-centered cubic where 4 atoms form an empty cubic cage.

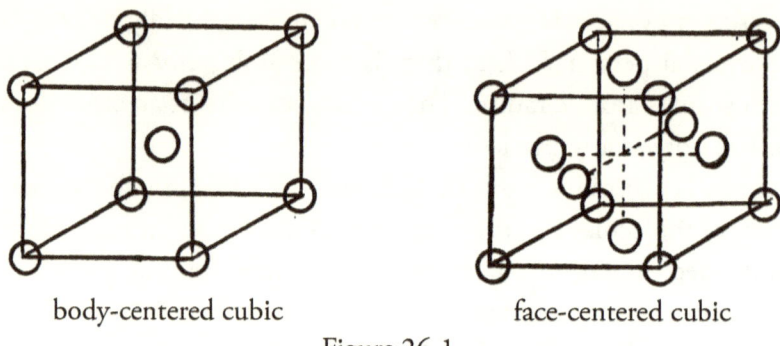

body-centered cubic   face-centered cubic

Figure 26-1

Aluminum has a face-centered cubic structure. Looking at one end of a face-centered cubic unit cell, it is apparent that the distance across the diagonal of the square formed by the nuclei of the four outer atoms is exactly four atomic radii. By the Pythagorean theorem, this is the square root of two times the distance along one side of the cube.

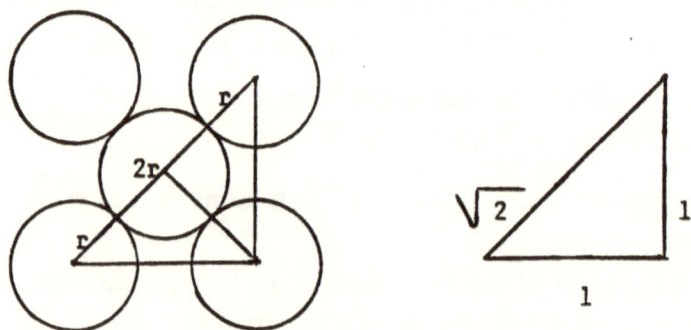

Figure 26-2

If a box is drawn from the nuclei of atoms forming the face-centered cube and the pieces of the atoms inside the box are counted, it turns out that the cube contains exactly 4 atoms. That is, 1/8 of each of the 8 corner atoms are contained inside the cube, and ½ of each of the 6 atoms in the faces of the cube are inside the box.

$$8 \left(\frac{1}{8} \text{ atoms}\right) = 1 \text{ atom}$$
$$6 \left(\frac{1}{2} \text{ atoms}\right) = 3 \text{ atoms}$$
$$\overline{}$$
$$4 \text{ atoms}$$

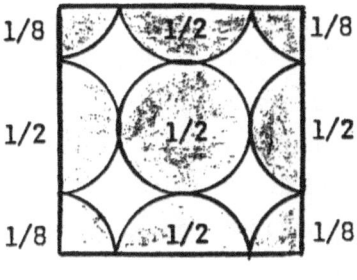

Figure 26-3

The cubes are called unit cells. One can calculate the volume of the unit cell when the density of the material is known. In the case of aluminum:

$$\left(\frac{4 \text{ atoms}}{\text{unit cell}}\right)\left(\frac{1 \text{ mole}}{6.022 \times 10^{23} \text{ atoms}}\right)\left(\frac{26.98 \text{ g}}{\text{mole}}\right)\left(\frac{1 \text{ cc}}{2.702 \text{ g}}\right) = \frac{6.63 \times 10^{-23} \text{ cc}}{\text{unit cell}}$$

The distance along one side of the unit cell is the cube root of the volume. The cube root of 6.63×10^{-23} or 66.3×10^{-24} is 4.05×10^{-8} cm. The 4 atomic radii of aluminum atoms along the diagonal of the unit cell, therefore, are 4.05×10^{-10} meter, according to the Pythagorean theorem. Thus, the radius of the aluminum atom calculated using interatomic spacings is $\frac{\sqrt{2}}{4} \times 4.05 \times 10^{10}$ meters = 1.43×10^{-10} meters.

The radius of an aluminum atom, according to the Bragg x-ray technique, is 1.25×10^{-10} meters.

In 1913, Niels Bohr found an alternate method for finding the atomic radius of the hydrogen atom. According to Bohr, the atomic diameter of a hydrogen atom should be equal to

$$\frac{\text{Planck's constant}}{(2\pi)(\text{electron mass})(\text{electron velocity})}$$

Bohr's calculation verified the calculation by the classical method.

Knowledge of atomic radii and interatomic distances turned out to be very important soon after they were discovered. In 1913, when Henry and Lawrence Bragg began using known interatomic distances to calculate the wavelengths of x-rays reflected off a crystal, they founded the field of x-ray spectroscopy. This technique has been used, and still is used, to find the molecular structures of complex molecules. This was also precisely the method that Henry Gwyn Jeffreys Moseley needed to find the number of protons in the nuclei of atoms and, consequently, to discover atomic numbers.

SIZES OF NUCLEI AND THE PROTON

The phenomenal rate of the growth of science during the early twentieth century is demonstrated by the fact that the approximate size of the nucleus of an atom—over a billion times smaller than one atom itself—was discovered the year after the size of the atom was discovered. In 1911, Lord Rutherford observed that about one out of 8,000 alpha particles was severely deflected while passing through a sheet of gold foil 2,000 atoms thick. Rutherford knew that the gold sheet was about 2,000 atoms thick because he knew the width of an atom due to the previous year's discoveries. He also knew the thickness of the foil from formulas

$$(\text{length})(\text{height})(\text{thickness of gold foil}) = \text{volume} \text{ AND}$$
$$\text{volume of foil} = \frac{\text{mass}}{\text{density }(19.3 \frac{g}{ml})}$$

Combining the equations, the thickness of the foil is

$$\left(\frac{\text{grams of gold}}{(\text{length of foil})(\text{height of foil})}\right) \frac{1 \text{ cc}}{19.3 \text{ gram gold}} = \text{thickness of foil}.$$

In terms of atoms, the thickness of the foil then is

$$\frac{\text{thickness of the foil}}{\text{diameter of a gold atom}} = \text{number of gold atoms thick}.$$

At that time, it was already known that the diameter of a gold atom is 2.68×10^{-10} meters.

These data can be used to estimate the cross-section area of a gold nucleus. If only one alpha particle in 8,000 was deflected by a foil 2,000 atoms thick, then only one alpha particle in $2,000 \times 8,000$ (i.e., 1.6×10^7) would be deflected if the foil were one atom in thickness. Evidently then, the cross section of a gold nucleus is about $(8,000)(2,000)$ smaller than the cross section of a gold atom. Given that, the cross section of a gold atom divided by the cross section of a gold nucleus should equal about 16,000,000 or 16×10^6.

$$\frac{\pi \,(\text{gold atom radius})^2}{\pi \,(\text{gold nucleus radius})^2} = 16 \times 10^6$$

Cancelling pi and taking the square root of both sides, we have

$$\frac{\text{radius of a gold atom}}{\text{radius of a gold nucleus}} = 4 \times 10^3$$

Plugging in the radius of a gold atom (1.34×10^{-10} meters),

$$\text{the radius of a gold nucleus} = \frac{1.34 \times 10^{-10} \text{ meters}}{4 \times 10^3} = 3 \times 10^{-14} \text{ meters}.$$

Modern techniques used to determine the size of a nucleus utilize high-speed electrons instead of alpha particles to bombard the nucleus, and the calculations reveal a slightly smaller nucleus size of approximately 7×10^{-15} meters. The problem with Rutherford's experiment was that the alpha particles had been deflected before they could impact, because they, like the nucleus, have positive charges. The modern technique is the same as Rutherford's, however, in that the scatter of the particles is observed, and the calculations are performed in the same way. A second technique, based on looking at the energies of x-rays produced when the orbit of a mu-meson around a nucleus decays, verifies the accuracy of the electron scatter technique.

Knowing the radius of a gold nucleus, the radius of a proton may also be determined. Since there are 197 nucleons (neutrons + protons) in a gold nucleus, the volume of the nucleus should be about 197 times the volume of a proton.

$$\tfrac{4}{3}\pi(\text{radius of a gold nucleus})^3 = 197 \times \tfrac{4}{3}\pi(\text{radius of a proton})^3$$

Cancelling, $\frac{4}{3}\pi$ we have

(radius of a gold nucleus)^3 = 197 (radius of a proton)^3

Taking the cube root of both sides, we have radius of a gold nucleus = 5.8 × radius of a proton, so

$$\text{radius of a proton} = \frac{7 \times 10^{-15} \text{ meters}}{5.8} = 1 \times 10^{-15} \text{meters}.$$

This is the approximate radius of the sphere of the proton's positive charge distribution.

ATOMIC NUMBERS

Atomic nuclei contain positively charged protons and (except for protium hydrogen) uncharged neutrons. Their combined masses plus the minimal mass of the surrounding electrons make up an atom's atomic mass. The characteristic that differentiates one element from another, however, is not the atomic mass but the number of positively charged protons in the nucleus. When Ernst Rutherford discovered in 1911 that an atom consists of a small, massive nucleus surrounded by an electron (cloud), no one knew how to find the nuclear charge.

It had been known for some years that when fast-moving electrons are suddenly stopped, electromagnetic radiation is given off. Faster electrons produce shorter wavelengths of radiation, and at high energies, x-rays are the form of radiation produced. In 1913, English physicist, Henry Gwyn Jeffreys Moseley, discovered that when electrons are impacted onto an anode in an evacuated chamber, the wavelength of x-rays produced is related to the atomic mass of the element from which the anode was constructed. With one or two exceptions, higher atomic mass anodes produce shorter wavelength x-rays. Guessing that the x-ray wavelength is inversely related to the number of protons in the nucleus (owing to the greater attraction of a more highly charged nucleus for the oppositely charged electron), Moseley found that $\frac{a\ constant}{(protons\ in\ the\ nucleus-1)^2}$ gives the shortest observed x-ray length.

Moseley's discovery solved a riddle that had plagued chemists since Mendeleev created the periodic table of elements. Based on the order of atomic masses, the element tellurium should have the characteristics of a halogen, while iodine should be almost metallic and have a valance of −2 when combined with a metal. In fact, the reverse is the case. Moseley's experiment showed that despite its higher atomic weight, tellurium actually has one less proton than iodine. In order of atomic number, iodine falls into its proper category as a halogen. It was also found that the elements cobalt and nickel were transposed—the lighter nickel having one more proton than cobalt.

It is not immediately obvious why x-ray wavelengths are inversely related to the square of the number of nuclear protons minus 1, rather than to the number of protons alone. The explanation given for this fact is that when an electron drops to the lowest orbital in an atom, that orbital is already occupied by another electron (two electrons fill the lowest orbital of any atom). While the electron is "dropping," it is attracted by all the protons in the nucleus and simultaneously repelled by the electron remaining in the lowest orbital. That electron effectively neutralizes the attractive power of one of the protons.

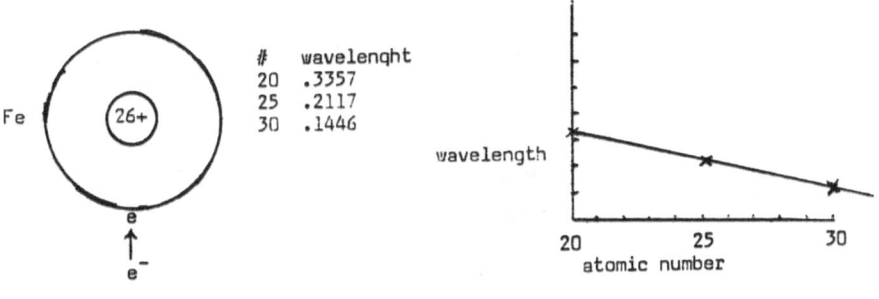

Figure 28-1

THE FINE STRUCTURE CONSTANT

The positions of the lines in the hydrogen spectrum wavelengths were explained with remarkable success by Johannes Rydberg's formula:

$$\frac{1}{\text{wavelength}} = (\text{Rydberg's constant}) \left(\frac{1}{\text{final orbit \#}}\right)\left(\frac{1}{\text{initial orbit \#}}\right)$$

Line positions are directly related to the amount of energy involved in an electron transition from one orbit to another. However, after the turn of the twentieth century, high-resolution spectroscopy began showing that the lines predicted by Rydberg's formula actually consisted of two or more lines very close together. This suggested that the quantum mechanical view at the time was too simplistic, even for the hydrogen atom.

The problem was first attacked by Arnold Sommerfeld who guessed that the fine-line splitting was due to energy variations of the electrons in their orbitals caused by special relativistic mass increase. When an electron is in a highly elliptical orbit, its velocity is high enough near the nucleus to increase its mass, slightly because of the equation

$$\text{electron's observed mass} = \frac{\text{electron's rest mass}}{\sqrt{1 - \frac{(\text{electron velocity})^2}{(\text{speed of light})^2}}}$$

Consequently, the electron will have slightly greater energy (kinetic energy being proportional to mass) near the nucleus. If such a fast electron changes orbital to another energy level, the energy change will be slightly different for an electron in a less elliptical orbit or an electron farther from the nucleus, making the same transition. Sommerfeld went on to develop an equation that correctly predicted the fine splitting in the hydrogen spectrum. The equation introduced a new constant *a* and is known as the fine-structure constant.

$$a = \frac{2\pi \text{(elementary charge)}^2}{\text{(Planck's constant)(the speed of light)}}$$

In 1925, two graduate students at Leiden University, Samuel Goudsmit and George Uhlenbeck, realized that Sommerfeld's calculations of the relativistic effects producing fine structure could not explain the multiple lines found in the spectra of group IA metals because some of the electrons responsible for the split spectral lines were moving at too low a velocity. Goudsmit and Uhlenbeck suggested that the fine structure might be the result of intrinsic spins of the electrons rather than from relativistic effects. The Goudsmit-Uhlenbeck theory proposed that if electrons had intrinsic spins, the fine structure could result from magnetic interactions inside the atom. An electron orbiting a nucleus sets up an atomic magnetic field because the electron's path acts like a tiny electric circuit.

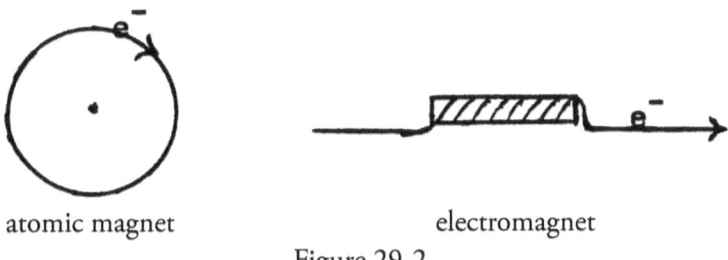

atomic magnet   electromagnet

Figure 29-2

An intrinsic spin or rotation of an electron would cause the particle to create its own magnetic field for the same reason. The magnetic field produced by the electron's spin is either aligned with or aligned against the magnetic field produced by its own orbit. When the atoms have more than one electron, each of the electrons contributes to the overall magnetic field of the atom. An electron, whose field is aligned with the overall magnetic field of the atom and which is undergoing an orbital transition, will require or give off a slightly different quantity of energy than an electron undergoing the same transition but aligned against the atom's magnetic field. This slight difference causes the splitting of the spectral lines.

In 1928, Paul Dirac showed that electron spin is a relativistic phenomenon, and that the displacement of the spectral lines in a one-electron atom is directly related to the square of the fine structure constant. Sommerfeld's relativistic calculations hadn't worked with elements other than hydrogen because they were based on mechanical considerations that fail when dealing with the submicroscopic. Dirac's rigorous relativistic treatment showed how all electrons have spins regardless of their speeds in atomic orbitals, and that the spins may have only one of two moments.

The fine-structure constant itself turns up in places besides fine-structure calculations. It is the unitless quantity that conveys the magnitude of interaction between a charged particle and an electromagnetic field. As a result, the constant turns out to be the ratio of the speed of an electron in the lowest orbit of a hydrogen atom to the speed of light (i.e., V/C = a). The present accepted value for the fine-structure constant is $7.29735257698 \times 10^{-3}$.

THE AGE OF THE EARTH

In 1654, Bishop James Ussher estimated the Earth's age was about 6,000 years based on chronologies he found in the Bible. Geologists were incredulous at this figure because, if it were true, geological processes would have had to occur much faster in the past than they presently do. Limestone formations, riverbed erosion, continental drift, salt formations from areas that were formerly seas, and chalk formations built of creatures that lived in those seas all suggest that the Earth is *very* old.

The first known attempt to discover the age of the Earth was made by Edmund Halley in the late seventeenth century. Lakes that have no outlets gradually became saltier over time because tributary streams continually bring in dissolved salts, but water that evaporates from the lakes doesn't carry them away. Halley reasoned that this phenomenon should also be happening to the oceans since they have no outlet. When Halley compared the estimated yearly salt addition to the oceans from rivers with the total salt content of the oceans, he arrived at a very great age. It was much greater than the 6,000 years, Ussher and many of his other contemporaries believed. Unfortunately, Halley hadn't considered that many areas that had once been ocean floors had risen up to become land and retained salt in the process. Consequently, even Halley's estimate for the Earth's age was far too low.

Currently, we have several dating systems, three of which are accurate to within a year. Using recent technology, yearly weather patterns are traceable back 9,000 years through tree ring analyses, 200,000 years through polar ice analysis, and 200,000 years through ocean floor sediment analysis. Unfortunately, because of the dynamic qualities of the Earth, these ultra-accurate dating methods fail beyond 200,000 years.

The only reliable known method capable of measuring extremely ancient things is called radioactive dating. The method was proposed by Bertram Boltwood, a coresearcher of Ernest Rutherford, as a technique for finding the age of rocks. Radioactive dating is done by comparing the amount of a radioactive element that was incorporated into the crystal when the rock was formed with the amount of the element left. The original amount is found by adding the amount of decay product that has formed over the years to the amount of the element that remains. Radioactive decay occurs in a regular way over long periods of time. It follows the law of decay, which states that the log of

$$\frac{\text{initial amount of radioactive element}}{\text{remaining amount of radioactive element}} = \text{a constant} \times \text{time}.$$

The constant depends on the identity of the radioactive element. Potassium-40 decays to argon-40 via beta decay or by electron capture. Potassium-40 has a half-life (half the original potassium decays) of 1.3 billion (milliard) years.

Rubidium-87 decays to strontium-87 via beta decay. The half-life of rubidium-87 is 47 billion years.

Uranium-238 decays to lead-206 through a series of short-lived decay products. The half-life for this process is 4.15 billion years.

The exceptionally slow decay rates for these elements make them useful in measuring the age of rocks. These dating systems are not perfect, however. Though it is fairly certain that no argon is present when a rock forms (argon is inert and doesn't form compounds with other elements) the potassium—argon dating system is flawed by the fact that argon is a gas and has a tendency to escape from more porous rocks. Hence, a rock dated by this system may appear younger than it actually is. There is also a problem with the uranium—lead dating system. When the rocks form, there is usually lead-206 as well as uranium present. This original lead-206 must be subtracted from the total lead-206 present. The amount to subtract is found by measuring the amount of lead-204. Lead-204 is not a decay product. It has been lead since its creation, presumably synthesized in the supernova that preceded our Sun. Furthermore, the ratio of primordial lead-204 to primordial (original) lead-206 (also created in the supernova) is fixed. Consequently, knowing the amount of lead-204 tells you how much lead-206 was in the rock in the first place. The remaining lead-206 is the decay product of uranium-238. Despite the problems with these techniques, they agree pretty closely with one another for almost all rocks, suggesting the problems are not exceptionally great.

When researchers began attempting to find the age of the Earth (about 1915), the facts that no existing rocks date back to the origin of the Earth, and that most rocks are geologically young, gave them fits. The oldest Earth rock ever found was in Australia and dates back over 3 1/2 billion (milliard) years. The problem of young rocks, however, has now been solved by dating meteorites. Analysis of isotope ratios of lead-206 and lead-207 suggest that both the earth and meteorites have a common origin. All meteorites tested (except two) show an age of between 4.2 and 4.6 billion years. The two exceptions are suspected of having been blasted out of the Moon in one case and Mars in the other.

THE SIZE AND AGE OF THE UNIVERSE

In view of the immensity of the universe in both size and age, it should not be surprising that these quantities are very difficult to ascertain. It is no less than remarkable that we should have any handle on these quantities at all, but in fact, we do.

When you are standing next to a railroad track and a train sounding its horn passes, you notice at the moment the train passes that the pitch of the horn suddenly lowers. The increase in frequency observed in approaching objects and the decrease in frequency by receding ones is common to all wave phenomena and is called the Doppler effect. Measuring the drop in frequency emitted by a receding object compared with the frequency emitted when it is not moving makes it possible to calculate how fast it is going away. In the case of sound, the equation is

$$\text{frequency heard} = \frac{\text{frequency emitted}}{1 + \frac{\text{the speed the object is receding}}{\text{the speed of sound}}}$$

The equation for light is

$$\text{frequency observed} = \text{frequency emitted} \left[\frac{1 - \frac{\text{the speed the object is receding}}{\text{the speed of light}}}{\sqrt{1 - \frac{\text{the speed the object is receding}}{\text{the speed of light}}}} \right]$$

In the 1920s, American astronomer Edwin Hubble observed that the frequencies of the characteristic line spectra of the elements were lower for distant galaxies than for nearby ones. The distances to the galaxies were estimated on the basis of the observed brightness of stars in them called Cepheid variables. These stars periodically vary in brightness. The star's period allows calculation of its absolute brightness, and knowing its absolute brightness allows its distance to be calculated using the inverse square law and its observed brightness. Line spectrum frequencies were determined using common spectrographic methods (spreading the light out with prisms or diffraction grating).

Not only were the distances to the galaxies inversely related to the line spectra frequencies, but when Hubble extrapolated the positions of the galaxies backwards, they all seemed to have their origins at the same point in space. It seemed as if the entire universe owed its existence to an explosion at that single point.

When Hubble was doing his research, there were difficulties in determining the distances to galaxies. They were thought to be generally closer than they are believed to be now. When Hubble reported his values in 1931 based on the idea that $\frac{\text{speed of galaxy}}{\text{distance to galaxy}} = \frac{\text{speed of light}}{\text{distance to edge to explosion (Universe)}}$ he obtained an age of 1.8 billion (milliard) years. This figure was absurd because it was already known that the Earth was older than that. Refinements in methods giving galaxy distances have given us an estimated age of the universe eight times that of Hubble's (13.8 billion years). As to the size of the universe, if it did originate in an explosion, it should have a radius of 13.8 billion light years.

Speculative as Hubble's idea of an expanding universe may sound, there is mounting evidence from both radio astronomy and particle physics that it is true. In 1965, radio telescopes turned to a frequency that corresponds to a black-body radiation peak emitted by objects with temperatures between two and three degrees above absolute zero and picked up signals from the sky—every part of the sky! The only explanation anyone has come up with is that energy is reaching us from the "shell" of the receding big bang. George Gamow had predicted the existence of this radiation in 1948. Recent theories on the nature of forces also seem to support the big bang. The electroweak theory that attempts to explain the relationship between electric and weak nuclear forces proposes that at a time in the distant past, shortly after the big bang, electric and weak forces were indistinguishable from one another. In the late twentieth century, experimental particle physics gave electroweak theory dramatic confirmation by finding the previously unknown particles it predicted, the W and Z particles.

THE TEMPERATURE OF THE EARTH'S CORE

As late as the turn of the twentieth century, many geologists believed that the interior of the Earth was relatively cool. Volcanoes were thought to form from pockets of magma. Conditions in deep mines, however, contradicted the cool interior idea, and it is now believed that the temperature increases nearly 5 degrees Celsius for every kilometer of descent down to 350 kilometers below the surface. That the Earth's interior is very hot is substantiated by the fact that experiments timing and measuring waves produced by earthquakes and nuclear bombs have established that the Earth's core, or at least the outer part of it, is liquid.

Earthquake waves, called S waves, are transmitted through solid but not liquid materials. The Earth's core does not transmit S waves. The high density of the liquid (recall the Cavendish calculation of the Earth's mass) indicates that it is probably constituted of materials with high melting points (e.g., iron, iron oxide, very dense rock, or some mixture of the three). Under the tremendous pressure of the Earth's interior, the melting points are even higher than the area at the surface, so the temperature of the core (at least the liquid part) has to be at least 2,000 K.

When one first considers the fact that the Earth's interior is hot, it seems strange. Why hasn't a 4 ½-billion-year-old earth radiated all of its inner heat out into space? It was precisely this question that led nineteenth century geologists to the conclusion that the center of the Earth was cool. But Henri Becquerel's discovery of radioactivity in 1896 provided an answer to the problem. The Earth's interior contains radioactive materials which decay, producing high-energy particles (alpha, beta, and gamma rays). These particles cannot penetrate the thousands of kilometers of rock and iron to escape into space. Instead, they strike and impart their energies to neighbor atoms. That energy ultimately ends up in the form of atomic or molecular motion (heat). Radiation occurs continually, so heat is continually produced.

Though it is known that the Earth's core is hot, it is not known precisely how hot. The temperature at the boundary of the Earth's inner and outer core is now estimated at approximately 6,000 C (10,800 F).

Scientist developed this estimate by determining the melting point of iron at high pressures. The inner/outer core boundary temperature, where pressure is 3.3 million times the pressure at sea level, is about 10,832 F plus or minus 930 degrees.

It seems strange that more is known about the temperatures of stars millions of light years away than is known about a part of our own planet just 3,000 kilometers away, but such is apparently the case. We will have to wait for the developed of a new instrument or some ingenious technique before the question of the temperature of the Earth's core can be answered.

BIBLIOGRAPHY

Abell, George. *Exploration of the Universe*. California: Harcourt College, 1995.

Boyer, Carl. *A History of Mathematic*. New Jersey: Wiley, 1991.

Bromwich, T. J. I. *An Introduction to the Theory of Infinite Series*. New York: Merchant Books, 2008.

Cajori, Florian. *A History of Mathematics*. South Carolina: Bibliolife, 2009.

Eisberg, Robert, and Robert Resnick. *Quantum Physics*. New Jersey: Wiley, 1985.

Friedlander, Michael. *Astronomy: From Stonehenge to Quasars*. New Jersey: Prentice Hall, 1985.

Frisinger, Howard H. *A History of Meteorology to 1800*. New York: American Meteorological Society. Science History Publications, 1977.

Greenburg, Leonard. *Physics with Modern Applications*. Pennsylvania: Saunders, 1978.

Halliday, David, and Robert Resnick. *Physics*. New Jersey: Wiley, 1978.

Hawthorne, Robert M., Jr. "Avogadro's Number. Early Values by Loschmidt and Others." *Journal of Chemical Education* 47 (November 1970): 751–754.

Jammer, Max. *The Conceptual Development of Quantum Mechanics*. New York: McGraw Hill, 1966.

Kohn, Moritz. "Josef Loschmidt (1821–1895)." *Journal of Chemical Education* 22 (August 1945): 381–384.

Leicester, Henry, and Herbert Klickstein. *A Source Book in Chemistry*. Massachusetts: Harvard University Press, 1952.

Leithold, Louis. *The Calculus 7*. New York: Harper Collins College Division, 1995.

MacDougall, F. H. *Physical Chemistry*. New York: MacMillan, 1937.

Magie, William Francis. *A Source Book in Physics*. Massachusetts: Harvard University Press, 1935.

McBrien, V. O. *Introductory Analysis*. New York: Appleton-Century Crofts, Inc., 1961.

McGrow Hill Encyclopedia of Science and Technology. 2012.

Merkin, Melvin. *Physical Science with Modern Applications*. Boston: Cengage Learning, 1993.

Milsner, Charles, Kip Thorne, and John Wheeler. *Gravitation*. New York: W. H. Freeman, 1973.

Moulton, Forest Ray. *An Introduction to Astronomy*. New York: MacMillan, 1925.

Pasachoff, Jay. *Astronomy from the Earth to the Universe*. Boston: Cengage Learning, 1997.

Petrik, Eugene. *Inductive Calculus*. Massachusetts: Allyn and Bacon, 1967.

Wells. David. *The Penguin Dictionary of Curious and Interesting Numbers*. New York: Penguin,1997.

Prutton, Carl, and Samuel Maron. *Principles of Physical Chemistry*. New York: MacMillan, 1965.

Ronan, Colin. *Edmund Halley*. New York: Doubleday, 1969.

Segre, Emilio. *Nuclei and Particles*. San Francisco: Benjamin Cummings, 1977.

Shapely, Harlow, and Helen Howarth. *Source Book in Astronomy*. New York: McGraw Hill, 1929.

Sienko, Michell, and Robert Plane. *Chemistry: Principles and Properties*. New York: McGraw Hill, 1966.

Stautberg, Margaret. *Physics, the Excitement of Discovery*. California: Wadsworth, 1983.

Thining, Hans. "Ludwig Boltzmann." *Journal of Chemical Education* 29 (June 1952): 298–299.

Tipler, Paul. *Modern Physics*. New York: W. H. Freeman, 2012.

Wells, David. "Ramanujuan and Pi." *Scientific American* 256 (February 1988): 112–117.

Note: Quark Theory, as it is presently accepted in science, involves not only quarks, but also "gluons" and "color charges" which change color. I believe that Quark Theory as it is presently accepted is wrong. It is too complicated. I believe that quarks are attracted to each other because they are semi-mirror images of each other. As far as I know, I am the only one who believes this.

It is customary in science, after discovering some kind of order or pattern, to form a hypothesis and test it by experiment. Because particle physics is not my area of expertise, I was unable to determine an appropriate experiment. But the pattern presented here is impressive in its implications for fundamental particle structures. By publishing the material, I hope to bring it to the attention of those more knowledgeable in this area, and that someone else may develop appropriate experiments to test my hypothesis.

SOME BACKGROUND

Electrons and protons, primary constituents of atoms, were both "discovered" (i.e., some of their properties were measured) in the late 1890s. The third primary constituent of atoms, the neutron, was discovered in 1932. After that, new particles not associated with atoms (at that time) began turning up. Some of them were predicted and some were not. One predicted particle was the neutrino, an uncharged invisible particle. The neutrino was predicted (in 1930) to account for apparently unconserved mass and momentum in radioactive decay. It was discovered in 1956.

Another type of particle that began turning up (starting with the positron in 1932) was the antiparticle (so-called antimatter). A particle's antiparticle is its mirror image. If a particle meets its antiparticle, both are destroyed and a photon (light) is produced.

As particles were discovered, they were put into one of three groups:

1. Particles and antiparticles that did not interact with strong nuclear force were called leptons.
2. Particles that interacted with strong nuclear forces and decayed to protons and other particles were called baryons.
3. Particles that interacted with strong nuclear force and decayed but did not produce protons as decay products were called mesons.

Among the mesons and baryons were particles that took unusually long times to decay. These were called strange particles. It was said that they had the property of strangeness.

In 1962, Murry Gell-Mann at Cal Tech found a relationship between the mesons and baryons (collectively called hadrons). He called the relationship the eightfold way. In 1964, he discovered that the eightfold way relationship was analogous to predicting that the meson and baryons were built up of other particles. Gell-Mann called these hypothetical particles quarks. The baryons were each composed of three quarks, while the mesons were each composed of a quark and an antiquark. In 1964, all known hadrons could be shown to be built up of just three kinds of quarks. Gell-Mann called these up, down, and sideways (strange). The strange particles each possessed one or more strange or antistrange quarks. Since that time, other particles have been discovered that were evidently made up of quarks other than those named above. The names of the "new" high-energy quarks are charm, top (or truth), and bottom (or beauty).

HYPOTHESIS

The picture of particle physics has been confused. There are too many elementary particles for them to be elementary. So let's make a supposition. Let's <u>suppose</u> that all elementary particles are themselves made up of a single kind of particle. Let's look and see if we can find a good candidate for this particle.

If we assume that all particles are built up of a single kind of elementary particle, then that particle must be the simplest and most elementary of all particles. This eliminates as candidates of all particles that decay because a particle's decay products have to be more elementary than the parent particle. Okay, so what particles do not decay?

- Photons (light)
- neutrinos
- antineutrinos
- electrons
- antielectrons (positrons)
- protons
- antiprotons

These are the only particles that do not decay. However, protons are apparently built of up quarks and a down quark, and antiprotons are apparently built of anti-up quarks and an antidown quark. So this changes our list of candidates to the following:

- photons
- neutrinos
- antineutrinos
- electrons
- positron
- up quarks
- down quarks,
- anti-up quarks
- antidown quarks

Now which of these particles is the simplest. Well, electrons have spin, charge, and mass. Quarks have spin, charge, mass, and strong force. Photons, neutrinos, and antineutrinos have spin only. (Note: Mass is now claimed, but there is reason to doubt this. The light and the neutrinos from Supernova 1987A arrived at the same time. If neutrinos had mass, they would have lagged behind. However, whether or not neutrinos and antineutrinos have mass, the conclusions of this paper remain the same.) Photons, neutrinos, and antineutrinos appear to be simpler than quarks, antiquarks, electrons, and positrons because they lack complexities like strong force. The photon, the neutrino, and the antineutrino appear to be the simplest known particles. But which is simpler, the neutrino or the photon?

When a muon (high-energy state electron) decays, it can decay into an electron and a photon or into an electron, neutrino, and antineutrino. Here it seems that the photon is somehow equivalent to a neutrino and antineutrino. Hence, the neutrino and its antiparticle, the antineutrino, appear to be the simplest known particles in the universe. If this supposition is wrong, we should be able to find ways to disprove it.

To build the universe out of neutrinos and antineutrinos, it will be necessary (for the time being) to ignore a phenomenon called conservation of charge, for which no exception has yet been found. During the development of physics, a number of quantities were observed to be conserved every time changes took place. These conserved quantities included mass, charge, momentum, energy, and parity (or symmetry). As time went on, it was discovered that some of these quantities were not conserved in all cases. It was discovered by Einstein (1907) that mass was not a conserved quantity because it could be converted to energy, and that energy was not a conserved quantity because it could be converted to mass. In 1956, it was discovered that in the radioactive decay of cobalt-60 to nickel-60, parity was not conserved. Now in order to build charged particles out of neutral particles (neutrinos and antineutrinos) it is necessary to assume that, as with mass, energy, and parity, there are certain circumstances in which charge is not conserved.

If we are successful in showing that other particles are composites of neutrinos and antineutrinos, then it should be possible to predict in exactly what circumstances charge should not be conserved and to verify them experimentally.

The simplest of particles other than the photon, neutrino, and antineutrino are (for the above stated reasons) the stable particles:

- electron
- positron
- up quark,
- down quark
- anti-up quark
- antidown quark

Assuming that particles are built up of neutrinos and antineutrinos in the simplest possible fashion, what are the simplest arrays of neutrinos and antineutrinos that could account for these particles?

(Note: Henceforth, antineutrinos will be abbreviated *a*, and neutrinos will be abbreviated *n*.)

From the muon decay, the photon structure may be abbreviated *an*. The other duet particle possibilities, *a-a* and *n-n* could account for only two particles. However, the simple particle arrays have to account for the six particles listed above. The next simplest particle arrays, after duets, are triplets. Triplets of neutrinos and antineutrinos hooked together in triangular fashion could account for four particles:

Again, however, there are not sufficient combinations. However, triplets of neutrinos and antineutrinos hooked together in linear combination do produce six possible particle structures:

a-a-a   n-n-n   a-n-a   n-a-n   a-a-n   n-n-a

Of these hypothetical structures, two are of an unmixed type: *a-a-a* and *n-n-n*. Four are mixtures of neutrinos and antineutrinos: *a-n-a*, *n-a-n*, *a-a-n*, and *n-n-a*.

Coincidentally, of the aforementioned particles, two are of the lepton type: electron and positron. Four are of the quark type: up, down, anti-up, and antidown. So let's suppose that the unmixed structures are the structures of the leptons, while the mixed structures are the structures of the quarks.

It was stated earlier that other high-energy quarks have been found (no quark has ever been isolated). High-energy leptons (electrons) have also been discovered. These new quarks have been assigned charges that correspond to charges of lower energy quarks. The high-energy (actually high mass) leptons likewise carry charges similar to their low energy counterparts. Let's assume that a high-energy quark that carries the same charges as a stable quark is simply a high energy form of that quark, and a high-mass electron that carries the same charge as the electron is just an excited state of the electron.

Now it is necessary to make one last assumption before testing the validity of all of them. A particle's anti-particle is its mirror image. Let's assume the same is true for the neutrino-antineutrino structures. So for *a-a-a*, the antiparticle is *n-n-n*. For *n-a-n*, the antiparticle is *a-n-a*. For *n-n-a*, the antiparticle is *a-a-n*.

Here is a list of suppositions that have to be made:

1. All particles are composites (built out of) the simplest particle and its antiparticle.
2. The simplest particle is the neutrino.
3. There are certain circumstances under which charge is not conserved.
4. The stablest particles other than the neutrino, the antineutrino, and the photon (i.e., particles that have mass and do not decay) are made up of neutrino and antineutrino triplets—the two leptons having unmixed structures, and the four quarks having mixed structures.
5. A high-energy quark that has the same assigned charge as a stable quark is just an excited state of that quark, and a heavy

electron that carries the same charge as an ordinary electron is just an excited state of that electron.
6. Mirror-image structures correspond to the mirror-image particles they represent (i.e., *a-a-a*—electron, *n-n-n*—electron OR *n-n-n*—positron, *a-a-a*—positron).

Assigned Charges of Quarks and Leptons

Assigned Charge	Stable State	High-Energy State	Higher Energy
-1	electron	muon	tau particle
+1	positron	positive muon	positive tau
+2/3	up	charm	top
-2/3	anti-up	anticharm	antitop
+1/3	antidown	antistrange	antibottom
-1/3	down	strange	bottom

We can check to see if the neutrino/antineutrino structure hypotheses is possible or not by looking at particle decays. We assume that both particles and their decay products are made of neutrinos and antineutrinos. So if, for example, a parent particle has two more neutrinos than antineutrinos in its structure, then its decay products must also have two more neutrinos than antineutrinos in their structures.

Let's look at the neutron decay. A neutron decays into a proton, an electron, and an antineutrino. Symbolically, it is N→P⁺ e⁻ + a. When structures were assigned to leptons and quarks, it was assumed that electrons were the unmixed structures (*a-a-a* or *n-n-n*). Let's see if it is possible that electrons are antineutrino triplets *a-a-a*.

A neutron (supposedly) is composed of two down quarks and one up quark, while a proton is composed of two up quarks and a down quark. So when a neutron decays to a proton, effectively, a down quark is converted into an up quark, while the other quarks remain unchanged.

ddu → duu + electron + anti-neutrino. If the electron is assumed to be the *a-a-a* triplet, it does not matter what mixed structures are assigned to the up quark and down quark, the decay is impossible.

	Down		up		electron		Antineutrino
1.	a-a-n	→	a-n-n	+	a-a-a	+	a
	1 neutrino		2 neutrinos				
	2 antineutrinos		5 antineutrinos				
OR							
2.	a-n-n	→	a-a-n	+	a-a-a	+	a
	1 neutrino		1 neutrino				
	2 antineutrinos		6 antineutrinos				

There are simply too many antineutrinos in the decay products. In case 1, there is one more antineutrino than neutrino in the parent particle, but there are three more antineutrinos than neutrinos in the decay particles. Case 2 is even more unbalanced. So we can conclude that it is impossible that the electron is an antineutrino triplet and that the neutrino structure hypothesis is true.

However, there were two possible unmixed structures for the electron, *a-a-a* and *n-n-n*. Let's see if the *n-n-n* electron structure runs into the same difficulty.

	down		up		electron		Antineutrino
1.	a-a-n	→	a-n-n	+	n-n-n	+	a
	1 neutrino		5 neutrinos				
	2 antineutrinos		2 antineutrinos				
OR							
2.	a-n-n	→	a-a-n	+	n-n-n	+	a
	2 neutrinos		4 neutrino				
	1 antineutrino		3 antineutrino				

Case 1 does not work. There is an excess of one antineutrino in the parent particle, while there is an excess of three neutrinos in the decay particles. However, case 2 does work. There is one more neutrino than antineutrino in the parent particle, and there is also one more neutrino than antineutrino in the decay products. Case 2 says that the neutrino structure hypothesis is possible if, and only if, the electron is a neutrino triplet, the down quark is composed of two neutrinos and one antineutrino, and the up quark is composed of one neutrino and two antineutrinos.

Is it simply an accident or coincidence that these structures work for neutron decay? If it is just an accident, then they probably will not work for some other decay (recall that the *a-a-a* electron structure did not work for neutron decay). It is nearly impossible that structures that accidentally work in neutron decay would also accidentally work in all particle decays. So let's look at other particle decays using the structures that worked for the neutron decay.

electron	Antielectron (positron)	down quark	up quark	antidown quark	anti-up quark
n-n-n	a-a-a	a-n-n OR n-a-n	a-a-n OR a-n-a	n-a-a OR a-n-a	n-n-a OR n-a-n

We will arbitrarily use the structure nan for the down quark and *a-n-n* for the up quark in the following examples:

Neutron		Proton			Antineutrino	
N	→	p⁺	+	e⁻	+	$\bar{v}e$
d d u	→	d u u	+	e⁻	+	$\bar{v}$
(n-a-n) (n-a-n) (n-a-a)	→	(n-a-n) (n-a-a) (n-a-a)	+	(n-n-n)	+	a

An excess of 1 neutrino → an excess of 1 neutrino.
(i.e., 4 antineutrinos + 5 neutrinos → 6 antineutrinos + 7 neutrinos)

Π^+	$\rightarrow$	μ_+	+	v_v
$\bar{d}$ u	$\rightarrow$	μ_+	+	v
(a-n-a) (n-a-a)	$\rightarrow$	(a-a-a)	+	n

Excess of 2 antineutrinos → excess of 2 antineutrinos.
(i.e., 4 antineutrinos + 2 neutrinos → 3 antineutrinos + 1 neutrino)

μ_+	$\rightarrow$	e^+	+	$\bar{v}e$	+	v_v
μ_+	$\rightarrow$	e^+	+	$\bar{v}$	+	v
(a-a-a)	$\rightarrow$	(a-a-a)	+	a	+	n

Excess of 3 antineutrinos → excess of 3 antineutrinos.
(i.e., 3 antineutrinos → 4 antineutrinos + 1 neutrino)

Π^-		$\rightarrow$	μ^-	+	$\bar{v}_v$
$\bar{u}$	d	$\rightarrow$	μ^-	+	$\bar{v}$
(a-n-n)	(n-a-n)	$\rightarrow$	(n-n-n)	+	a

Excess of 2 neutrinos → excess of 2 neutrinos.

μ^-	$\rightarrow$	e^-	+	$\bar{v}e$	+	v_v
μ^-	$\rightarrow$	e^-	+	$\overline{+v}$	+	+v
(n-n-n)	$\rightarrow$	(n-n-n)	+	a	+	+n

Excess of 3 neutrinos → excess of 3 neutrinos.

$\Pi°$		$\rightarrow$	γ	+	γ
$\bar{u}$	u	$\rightarrow$	γ	+	γ
(a-n-n)	(n-a-a)	$\rightarrow$	(a-n)	+	(a-n)

Excess of zero neutrinos → excess of zero neutrinos.
(i.e., 3 neutrinos + 3 antineutrinos → 2 neutrinos + 2 antineutrinos)

Σ°			→	N			+	γ
u	d	s	→	u	d	s	+	γ
(n-a-a)	(n-a-n)	(n-a-n)	→	(n-a-a)	(n-a-n)	(n-a-n)	+	(a-n)

Excess of 1 neutrino → excess of 1 neutrino.

Σ⁻			→	N			+	Π⁻	
d	d	s	→	u	d	d	+	ū	d
(n-a-n)	(n-a-n)	(n-a-n)	→	(n-a-a)	(n-a-n)	(n-a-n)	+	(a-n-n)	(n-a-n)

An excess of 3 neutrinos → an excess of 3 neutrinos.

Ξ°			→	Λ°			+	Π°	
u	s	s	→	u	d	s	+	ū	u
			→	(n-a-a)	(n-a-n)	(n-a-n)	+	(a-n-n)	(a-a-n)

Excess of 1 neutrino →excess of 1 neutrino.

Ξ⁻			→	Λ°			+	Π⁻	
d	s	s	→	u	d	s	+	ū	d
(n-a-n)	(n-a-n)	(n-a-n)	→	(n-a-a)	(n-a-n)	(n-a-n)	+	(a-n-n)	(n-a-n)

Excess of 3 neutrinos → excess of 3 neutrinos.

Ω⁻			→	Ξ⁻			+	Π⁻	
s	s	s	→	u	s	s	+	ū	d
(n-a-n)	(n-a-n)	(n-a-n)	→	(n-a-a)	(n-a-n)	(n-a-n)	+	(a-n-n)	(n-a-n)

Excess of 3 neutrinos → excess of 3 neutrinos.

K°		→	γ	+	γ
d̄	s	→	γ	+	γ
(a-n-a)	(n-a-n)	→	(a-n)	+	(a-n)

Excess of zero neutrinos → excess of zero neutrinos.
(i.e., 3 neutrinos + 3 antineutrinos → 2 neutrinos + 2 antineutrinos)

K°		→	Π⁺	+	e⁻	+	$\overline{v_e}$
d̄	s	→	d̄ u	+	e⁻	+	v̄
(a-n-a)	(n-a-n)	→	(a-n-a) (n-a-a)	+	(n-n-n)	+	a

Excess of zero neutrinos → excess of zero neutrinos.
(i.e., 3 neutrinos + 3 antineutrinos → 5 neutrinos + 5 antineutrinos)

Λ°			→	p⁺			+	Π⁻	
u	d	s	→	u	u	d	+	ū	d
(n-a-a)	(n-a-n)	(n-a-n)	→	(n-a-a)	(n-a-a)	(n-a-n)	+	(a-n-n)	(n-a-n)

Excess of 1 neutrino → excess of 1 neutrino.
(i.e., 5 neutrinos + 4 antineutrinos → 8 neutrinos + 7 antineutrinos)

Σ⁺			→	N			+	Π⁻	
u	u	s	→	u	d	d	+	d̄	u
(n-a-a)	(n-a-a)	(n-a-n)	→	(n-a-a)	(n-a-n)	(n-a-n)	+	(a-n-a)	(n-a-a)

Excess of 1 antineutrino → excess of 1 antineutrino.
(i.e., 4 neutrinos + 5 antineutrinos → 7 neutrinos + 8 antineutrinos)

K⁺		→	μ⁺	+	v_v
u	s̄	→	μ⁺	+	v
(n-a-a)	(a-n-a)	→	(a-a-a)	+	n

Excess of 2 antineutrinos → excess of 2 antineutrinos.

D⁰		→	K⁻		+	Π⁺		+	Π⁰	
c	ū	→	ū	s	+	u	d̄	+	u	ū
(n-a-a)	(a-n-n)	→	(a-n-n)	(n-a-n)	+	(n-a-a)	(a-n-a)	+	(n-a-a)	(a-n-n)

Excess of zero neutrinos → excess of zero neutrinos.

D⁺		→	K⁻		+	Π⁺		+	Π⁺	
c	d̄	→	ū	s	+	u	d̄	+	u	d̄
(n-a-a)	(a-n-a)	→	(a-n-n)	(n-a-n)	+	(n-a-a)	(a-n-a)	+	(n-a-a)	(a-n-a)

Excess of 2 antineutrinos → excess of 2 antineutrinos.

K⁰		→	Π⁺		+	Π⁻	
d̄	s	→	d̄	u	+	ū	d
(a-n-a)	(n-a-n)	→	(a-n-a)	(n-a-a)	+	(a-n-n)	(n-a-n)

Excess of zero neutrinos → excess of zero neutrinos.
(i.e., 3 neutrinos + 3 antineutrinos → 6 neutrinos + 6 antineutrinos)

It seems unlikely that the structure assignment that worked for the neutron decay would coincidentally work for the decays of all particles if there weren't something valid about the structure assignment. A true test of the hypothesis would be to find an example of charge not being conserved. The best experimental candidate, I believe, would be to show that under some condition (currently unknown) a particle with equal numbers of neutrinos and antineutrinos could decay completely into photons (the charge disappearing in the process). The possible candidates for this experiment are the deuteron, the alpha particle, carbon nuclei, nitrogen nuclei, and oxygen nuclei. As to what the high-energy states of quarks and leptons might be, the fact should be considered that three atom molecules have several quantized high-energy states (spin, bend, symmetric stretch, and antisymmetric stretch).

MECHANISMS OF ELECTRIC FORCE

As in previous papers, in this paper, neutrinos are abbreviated n and antineutrinos are abbreviated a.

One thing that is clear about the neutrino-antineutrino structure theory of particles is that the electric charge of the particle corresponds to the longest consistent end of a triplet composed of neutrinos and antineutrinos.

n a n	down quark	1 neutrino	-1/3 charge
n a a	up quark	2 antineutrino	+2/3 charge
a n a	antidown quark	1 antineutrino	+1/3 charge
a n n	anti-up quark	2 neutrinos	-2/3 charge
n n n	electron	3 neutrinos	-3/3 = -1 charge
a a a	positron	3 antineutrinos	+3/3 = +1 charge

In the world outside the nucleus, there are only two electron charges: +1 or *a-a-a* and -1 or *n-n-n*. If electric attraction or repulsion happens as a result of overlap of two particles, and only end-to-end overlap is allowed, then one can see how attractive virtual photons (*a-n*) are produced by opposite charge overlaps and repelling virtual photons are produced when there are *n-n* or *a-a* overlaps.

As stated in an earlier paper, mirrorimage particles are not stable together, but semimirror image particles are stable. Consequently, mirrorimage electrons and positrons destroy each other, while semi mirror image particles (e.g., protons and electrons that mirror each other in charge but not in mass) form hydrogen atoms.

If the mirror type of interaction results from end-to-end overlap, then

n n n | n a n	electron down quark	1/3 repel
n n n | a n a	electron antidown quark	1/3 attract 1 virtual photon (a-n)
n n n | | a a n	electron up quark	2/3 attract 2 virtual photons
n n n | | n n a	electron anti-up quark	2/3 repel
n n n | | | a a a	electron positron	3/3 attract 3 virtual photons

The situation is similar but reversed when positrons interact with quarks.

Something else can be seen with the mirror image interaction idea. This is that the quarks can all have end-to-end attractive interactions with themselves:

a a n | n-a-a	up up
n-a-n | | n-a-n	down down
n-a-n | a-a-n	down up

These semimirror interactions could be the source of the strong nuclear force. Here, the virtual photons formed presently go by another name—gluons.

STRONG FORCE

It is universally accepted that attractive forces between particles occur as a result of another particle (photon, meson, boson, gluon) exchanged between the other two. But that idea is counterintuitive because, unless the exchange particle has negative mass, it should impart momentum, pushing the attracting particles apart.

The largest example of an exchange particle involved in attraction is the hydrogen bonding in chemistry. In this case, a proton is exchanged between negatively charged parts of two molecules. In this case, the exchange of the proton has nothing to do with the attraction. The attraction is due to the electrostatic attraction of the proton for the electronegative parts of each molecule. The proton exchange is simply a result of that attraction. Could it be that in the other cases of attraction between particles that the particle exchange is also a result rather than the cause of attraction?

My proposal is that particles attract each other by some mechanism in which they seek their mirror images.

Particles that are perfect mirror images (e.g., electron-position) always destroy each other. But particles that are almost like mirror images form stable composites. An electron and proton are mirror images in charge but not in structures. Their composite is the hydrogen atom.

With respect to strong force, a neutron and a proton are also semimirror images. They are semimirror images in structure (the proton is 2 up quarks and a down quark, while the neutron is 2 down quarks and an up quark.

U	D
U	D
D	U

This is a mirror image.

They are also semimirror images in charge. The up quark has the opposite <u>but not equal</u> charge of the down quark. The stable component of the imperfect mirror image of the proton and the neutron is the deuteron, and in fact, all atomic nuclei except proton—hydrogen.

Now according to existing theory, all nucleons (protons and neutrons) will feel strong force for each other once electrostatic repulsion is overcome. Larger nuclei are created when free neutrons stick to an existing nucleus because of strong force. Then that nucleus undergoes beta decay, converting the neutron to a proton. But neutrons always stick to the structures that contain their semimirror image protons. Even though neutrons should stick together because of the strong force, you never hear of composites of pure neutrons (except in neutron stars in which neutrons are held together by gravity).

It seems reasonable that the composites (nuclei) of pure neutrons don't occur in nature because neutrons are not semimirror images of each other, and composites of protons and neutrons are common in nature because they are imperfect mirror images of each other.

In fact, it seems that stable structures in nature are either composites of semimirror image particles (atoms and nuclei) or one of the components of them (protons, electrons, quarks). Even a neutron could be considered a composite of semimirror image particles—an electron and a proton modified by an antineutrino (i.e., a neutron decays to an electron, proton, and antineutrino).

THE APPARENT SOURCE OF GRAVITATIONAL AND ELECTRICAL FIELDS

A charged particle "couples" its electric potential energy with other charged particles of the opposite sign. The amount of that charged particle's energy that is "coupled" with the other charged particles varies with its distance from them. When two particles of opposite signs are a great distance from each other, very little of the charge potential energy is involved with the potential energy of the other particle.

At a closer distance, say when an electron is captured by a proton, part of the potential energy between the two particles is destroyed and leaves as a photon of light with an energy equivalent to 13.6 volts, while the particle pair (now a hydrogen atom) loses (or has lost) mass equal to the mass equivalent of the photon's energy.

At an even closer distance, when an electron interacts with its antiparticle—a positron—all of the potential energy between the particles is involved, and photons produced by the interaction have the energy of the whole mass equivalent of the particle pair, while the masses of the particles disappear entirely.

Now if at any time only some of a particle's potential energy is interacting with other charged particles, depending on its distance from them, what is the rest of the particle's potential energy doing? Could it be interacting with the rest of the universe? When an electron and proton combine, they become closer together and interact with one another more strongly, but at the same time, they lose the mass equivalent of 13.6 volts. Having the reduced mass, they interact with the rest of the universe less strongly. In the case of the electron and positron, the electrical interaction between the two is total, and the mass of the pair (i.e., gravitational interaction with the rest of the universe) disappears completely.

A particle's interactions are divided between electrical interactions with other electric particles and gravitational interactions with the rest of the universe. Neutral particles (e.g., a neutron) interact only gravitationally.

When the electrical particles get closer together, the number of their electrical interactions increases. Interactions that had been gravitational now become electrical, and so the particle system now has less mass because of the reduced number of gravitational interactions.

The photon produced when this happens, as when an electron combines with a proton, does <u>not</u> carry away the extra mass. That photon can produce more mass (i.e., its mass equivalent) by virtue of the fact that it can separate two charged particles. In doing this, the two particles have fewer electrical interactions between one another. Some of the interactions that had been electrical are now gravitational and the particle system gains the mass equivalent of the photon as a result of that.

DE BROGLIE DERIVATION FROM PHYSICAL MODEL

What is the nature of a particle's interaction with another particle or its interaction with the rest of the universe? It can be shown that the de Broglie equation can be derived on the basis that a particle exists in equilibrium with a virtual photon that has an energy equal to the mass equivalent energy of the particle. That is, the particle is in equilibrium with its own virtual photon.

The diffraction patterns of particles passing through slits or double slits are predicted by the de Broglie equation.

Suppose that a particle is in equilibrium with its energy (or standing wave) form. That is, the particle exists part of the time as its own virtual photon, and the photon has the energy equivalent of the particle's mass. The wavelength of the photon is

$$\lambda = \frac{c}{v}$$

$E = hv$ so $\lambda = \frac{ch}{E}$

$E = mc^2$ so $\lambda = \frac{ch}{mc^2} = \frac{h}{mc}$ and $\frac{1}{\lambda} = \frac{mc}{h}$

Say this particle (and its virtual photon) pass through a diffraction grating. At what rate do the wave crests of the virtual photon pass through the diffraction grating?

The rate at which wave crests (cycles) pass by a given point in space is $\frac{\text{meter}}{\text{sec}} \times \frac{\text{cycles}}{\text{meter}} = \frac{\text{cycles}}{\text{sec}}$ so $V \times \frac{1}{\lambda}$ = frequency of wave crests.

Combining with the relation $\frac{1}{\lambda} = \frac{mc}{h}$ for the standing wave, the frequency of wave crests = $\frac{Vmc}{h}$

This would be the matter wave frequency. Now what is the apparent wavelength of this matter wave frequency?

$$v = \frac{c}{\lambda} = \frac{Vmc}{h} \text{ so } \lambda = \frac{h}{mV}$$

Can this virtual photon be not only the entity that interacts with other nearby charged particles but also with the rest of the universe? I think it can be <u>quantitatively</u> demonstrated that this is what happens.

In general relativity, a body's or particle's inertial mass is equivalent to its gravitational mass. In special relativity, a fast-moving particle has more inertial mass than it has when it is not moving.

If we accept general relativity, we have to conclude that a fast particle also has more gravitational potential energy with the earth than it has when it is not moving. It can be shown that the equation that predicts quantitatively how much a particle's mass increases from its velocity can be derived from the idea that the particle's observed mass actually comes from its rest mass, added to the mass equivalent of the energy of its velocity.

This, in turn, can be shown by a derivation assuming that the particle is in equilibrium with its own mass equivalent virtual photon, and that there is a Doppler blue shift in the frequency (hence energy and mass) of that virtual photon when the particle is moving rapidly.

WHY WE ARE COMPOSED MOSTLY OF MATTER RATHER THAN ANTIMATTER

Why we have mostly matter in our area of the universe rather than an equal mixture of matter and antimatter is a big question. I believe that there are two unwarranted assumptions that create confusion regarding this subject:

1. *The assumption that electrons, quarks, and photons are fundamental particles.* This is wrong. Electrons are linear neutrino

triplets, up quarks are linear triplets of 2 antineutrinos and 1 neutrino. Down quarks are linear triplets of 2 neutrinos and 1 antineutrino. The photon is a neutrino-antineutrino duet.

2. *The assumption that time didn't exist before the big bang.* The big bang came from a *spinning* (implying time) singularity from a black hole in a previous universe that either became too massive or too small or spinning too rapidly to be stable. That one (singularity) particle suddenly became many spinning particles. Conservation of angular momentum requires that most of the particles shot from one direction perpendicular from the spin access of the singularity would be ejected with a right-hand screw spin (antineutrinos) while most of the particles ejected in the opposite direction would have left-hand screw spin (neutrinos).

Given the structure assignments (above) of quarks and electrons, our area of the universe has many more neutrinos (left-hand screw particles) than antineutrinos.

The opposite side of the universe had been built mostly of antineutrinos which would coalesce mostly into antimatter right after the big bang.

ABOUT THE AUTHOR

The author was a mediocre student in college. It was because of those problems in school that he decided to try to really learn and understand those concepts that had seemed vague in college.

www.ingramcontent.com/pod-product-compliance
Lightning Source LLC
Chambersburg PA
CBHW030839180526
45163CB00004B/1379